I0056711

# GALILEAN VARIANCE

## THE REBIRTH OF CLASSICAL PHYSICS:
### A COHERENT TREATISE ON
### TIME, LIGHT & GRAVITY

**Book 1 of 4**
**LIGHT**

Copyright © 2025
Jason Verbelli

**ISBN** 978-1-969175-02-2
First Printing – September 2025

**Publisher**:
Magnevelli, LLC
6977 Navajo Rd., Suite 158
San Diego CA, 92119
Email: jason@galileanvariance.com
Website: www.galileanvariance.com

**Paperback Version**
**Printed in the USA**

**Thanks, and motivation dedicated to:**

Dr. Edward Henry Dowdye Jr.
Dr. Pierre-Marie Robitaille
Professor John Roy Robert Searl
Fernando D. Morris
D.M.
T.P.I.T.S.
General Virgil W. Banning
and of course, my Mom, Bonnie Verbelli

**Potential Vindication and Reinforcement for:**

Sir Isaac Newton, Christiaan Huygens, Robert Hooke, Giordano Bruno,

Johann Goethe, Nicolas Fatio, Georges LeSage, Robert Hooke,

Siméon Denis Poisson, Dr. Edward Dowdye, Dr. Pierre-Marie Robitaille,

Petr Beckmann, Ronald Hatch, Albert Abraham Michaelson, Paul Gerber,

Nikola Tesla, Oliver Heaviside, Dr. Louis Essen, Walter Bowman Russell,

Kristian Birkeland, Dr. Winston Bostick, Max Abraham, Ernst Gehrcke,

Stephen Crothers, David Michalets  Alexander Unzicker, Hannes Alfvén,

Professor John Searl, Sir William Rowan Hamilton, Halton Arp, Sir Fred Hoyle,

Dewey B. Larson, Eric Lerner, Dr. Anthony Peratt, Pari Spolter, Paul Marmet,

Dr. Ricardo Carezani, Tom Van Flandern, Harold Aspden, Howard C. Hayden,

Carl A. Zapffe, Milan R. Pavlovic, David Bergman, Dr. George Galeczki,

James Paul Wesley, Michael Disney, Dr. R.C. Gupta, Bibhas De, Dr. Marcel Pages,

Dr. Charles Bruce, Dr. Willard H. Bennett, Dr. Emil Wolf, John Stuart Reid,

Bruce DePalma, Ruggero Maria Santilli, Glenn Jessome, Jaim Harlow, Josh Toms,

Robert Otay, Matt Presti, Clay Taylor, Thomas Joseph Brown & many more.

This publication is not a substitute for these other books, but rather
I think it compliments them; I encourage people to also read:

*Discourses & Mathematical Illustrations pertaining to the Extinction Shift
Principle under the Electrodynamics of Galilean Transformations*
by, Dr. Edward Henry Dowdye Jr.

*Einstein Plus Two*
by, Petr Beckmann

*Escape from Einstein*
by, Ronald R. Hatch

*Optical Sciences - Solid State Laser Engineering*
by, Walter Koechner

Ⓜ
**M**agnevelli, LLC

# Table of Contents

— CHAPTER 1 —

# Getting Back to the Basics

Language, Semantics and Terminology matter. We mustn't lose our way in science.
...or confuse the meanings of the following:

• **Proof** = A Mathematical derivation to show how you arrive at the solution to an equation. Proof is a route to show how you got a mathematical answer.

• **Evidence** = Information and data collected through experimentation and observation.

• **Hypothesis** = A hypothetical scenario that is proposed.

• **Theory** = An explanation of a hypothetical scenario based within the constraints and mathematical consequences/ proofs of a given model.

Experiments and observations never prove a theory.
Experiments gather evidence to suggest an understanding is correct.
Mathematical proofs are derived to explain a scenario under a given framework.
When a new framework is presented, all past evidence is reinterpreted under that new framework to disprove a previous "understanding". The proofs under the new framework justify adopting a new mathematical procedure. A new theory is developed, abiding by the consequences of those constraints imposed by that model. The extraneous method is abandoned, and the simpler one is adopted.  Well, maybe in a perfect world anyway.

*"Today's scientists have substituted mathematics for experiments, and they wander off through equation after equation, and eventually build a structure which has no relation to reality."*
— Nikola Tesla

Some things never change, eh Mr. Tesla? Nikola Tesla was disgusted with the adoption of Lorentz invariance and the abandonment of Galilean invariance. Galilean invariance assumed the velocity of light stays the same no matter what. The reasoning for that is that Galilean invariance assumed there was an aether medium between planets which deformed as an excuse for the velocity of light always being the same. Lorentz invariance assumes there is a 4th dimensional space-time medium between planets which deforms as an excuse for the speed of light always being the same. No more excuses. This book sorts the wheat from the chaff. The wheat being simply that a shift in frequency IS a change in the velocity of light. The chaff is any invariant model of light, both relativity and aether theory.

We can reword Tesla's quote to say, "Today's scientists have substituted computer simulations for observation, and they wander off through animation after animation, and eventually build a Hollywood fantasy that has no relation to reality." Or that, "Today's scientists have substituted Lorentz invariant transformations for Galilean variant transformations, and they wander off though dark matter, black holes, and big bangs, building universities with archives and libraries full of books that have no relation to reality."

The arguments are succinctly laid out for people to scrutinize. The collapse of Lorentz invariant transformations by Dr. Edward Dowdye's *Extinction Shift Principle* is as revolutionary as the collapse of the geocentric model by Galileo Galilei. The replacement of Lorentz invariance is devastating to the established status quo. However, the collapse of relativity is quite encouraging for the newcomer of science and to all those able to unlearn and re-learn. There is another 100 years of discovery to be made from returning to a classical approach, but under Galilean *variance*.

Back to the drawing board we go. Back to the basics to learn proper fundamentals, before the invention of relativity... before the invention of QED... before string theory and a clear path to return to classical physics. Time to clean the filter through which we interpret reality. It's been a while, and the filter of Lorentz invariance/relativity is plagued with deadly black mold.

This is a revolution in the understanding of re-emissions, frames of reference and correctly applying fundamentals of optics and a varying speed of light within 3D Euclidean space. The contents of this book are a rebirth of classical physics under the framework of reformulated Galilean variant transformations. So, again, I welcome you to a new era of what I call 22nd Century Science. Let's skip ahead a generation, shall we?

Fig. 1.1

If Beethoven could make beautiful music even though he was deaf... then maybe I can make beautiful logic of science even though I'm just the village idiot. Stick around for the symphony. It will either be nails on a chalkboard or a melodious harmony. Either way, I hope it makes you shed a tear.

# Science vs Psyence

Science is a journey to acquire knowledge through experience using a methodical process of trial and error. But psyence is just messing with your mind. It's deviously using statistics and interpreting evidence with confirmation bias. Illusions of observation.

In this book, I make the argument that Galilean variance and Dr. Edward Dowdye's *Extinction Shift Principle* qualifies as science while relativity and its foundation of Lorentz invariance is circuitous psyence and "mathemagic." No need for conjuring a mathematical artifice as an excuse to sweep a paradox under the rug of cop out on an explanation under the guise that it's "too quantum" to grasp, or that "only PhD holding professionals have the right to talk about these topics."

I don't defer critical thinking to ChatGPT or have experts do my thinking for me. I listen to everyone, compare, rationalize and naturally lean towards what makes most sense. My heretical statements abide by the rules and consequences of the mathematical foundation of Galilean variance. Getting rid of contradictions of the invariant frameworks and merging the similarities.

Math is a language and a tool - but it has been deviously misused as 'wile'.
You can derive equations for any given hypothetical scenario. But not all scenarios or computer simulations are accurate, and many conflict. People normally think of math as an end-all-be-all validation for an argument. If an equation can be written for a scenario... that scenario MUST be accurate, right? Not necessarily at all. Specific examples will be given later. The mathematical proofs derived from Galilean variant transformations are mutually exclusive and diametrically opposed to those from Relativity's Lorentzian invariant transformations.

Relativity's Lorentz invariance MUST abide by the proofs/constraints provided by the math of that specific model which dictates the following:

• Mass literally increases with acceleration
• That gravity is not a force
• That time literally dilates
• And the velocity of light remains the same to *all* observers in **ALL** frames of reference.

There are quite a few more fallacies that Lorentz invariant transformations have managed to bring to "life," if you want to call it that. The product of relativity's invariant model of light has created a homunculus and a grotesque abomination like Frankenstein's monster. It's been wreaking havoc on the scientific community for over 100 years.

The following chapters shed light on alternatives from the opposing framework. And no, this book is not "above your pay grade." Give yourself a promotion and a raise then... you deserve it after dealing with the last 120 years of nonsense. Make Isaac Newton proud.

Isaac NEWTon

Fig. 1.2

# Evidence vs Proof

The scientific community has built a tower of Babel, confusing the world by inappropriately interchanging terminologies like "evidence" and "proof". So now, each term and phrase could have a different meaning depending on who uses it.

Proof in a court of law is not the same thing as proof in science. Proof in a court of law is a legalese term referring to a given "Exhibit A" or piece of evidence attempting to validate an accusation. If there is irrefutable video footage of someone committing a crime... that video evidence is submitted in court and is used to "prove" the accusation that the person committed the crime or not.

But in science, proof is simply math. Video footage provides evidence. The words "validation" and 'proof' are not the same thing in science. The current established dogmatic regime uses Lorentz invariance to justify their explanations/theories of how the universe operates.

The sentiment seems to be that if someone claims to have an opposing mathematical proof to justify interpreting the same scenarios and evidence... and new interpretations contradict the established dogmatic regime... that someone will automatically be ostracized and labeled "a crackpot", "fringe" or "a pseudoscientist".

The gatekeepers of established dogma insist their one mathematical framework under Lorentz invariant transformations is the end-all-be-all for interpreting the totality of the universe... From *ALL* frames of reference! And with every experiment, they say they "prove themselves right" over and over within their own idiocratic echo chamber. But, thanks to Dr. Edward Dowdye, the framework of Galilean variance can now yield the same predictions and solutions as ALL relativistic procedures. The same evidence over the last 100 years re-interpreted to DISPROVE Relativity. Galilean variance gives justification to abandon Relativity and adopt a completely different model for time, light, gravity and how we perceive reality itself.

# Credible Opposing Framework

Under Galilean variance and Dr. Dowdye's emission's theory:

• Time is constant and does **NOT** literally dilate.

• The velocity of light **IS** dependent upon the velocity of the light source.

• A change in relative frequency **IS** a change in the velocity of light from a new F.O.R.

• Mass does **NOT** literally increase with acceleration.

• Gravity **IS** a force and an emission.

• There is **NO** requirement for a constancy of light in **ALL** F.O.R. (frames of reference).

• There is **NO** requirement for **ANY** medium to propagate light or gravity.

• There is **NO** Aether, **NO** 4th Dimension & **NO** Space-Time.

Dr. Dowdye's Effectivity and Galilean variance is in opposition to Einstein's Relativity. The consequences of Galilean variant transformations being able to yield ALL the same predictions and solutions as Lorentz invariant transformations are devastating to Relativity. There are two mutually exclusive frameworks of mathematical proofs that can derive the same conclusions with completely different interpretations of the same evidence. Here are some comparisons of mathematical proofs under the framework of Galilean Transformations, rather than Lorentz Transformations like Relativity uses.

Fig. 1.3

Some differences...

Relativity under Lorentz invariance says time is relative:

$$t' \neq t$$

(Time is **not** the same from all frames of reference)

and
$$t = \frac{t_0}{\sqrt{1-\frac{v^2}{c^2}}}$$

(Time literally dilates)

but Classical Physics/Galilean variance says time is **not** relative:

$$t' = t$$

(Time **is** the same from **all** frames of reference)

And
$$\tau_{tr} = \frac{\tau_0}{\sqrt{1-\frac{v^2}{c^2}}}$$

(A "*transverse relative time shift*" is the mathematical equivalent to relativistic time dilation. The difference between clocks is not indicative of reality/time itself dilating.)

In the context of comparing relativistic methods to a classical route, the usage of the Greek symbol Tau with the subscript tr represents "proper time" or a specific characteristic time in classical frameworks. Those original equations involving transformations that adhere to so-called 'Galilean Relativity' are under Galilean INVARIANCE. Meaning the speed of light stays the same no matter what because of a deforming aether medium. That is not to be confused with the modified derivations provided by Dr. Edward Dowdye. Showing that the velocity of light is VARIABLE, meaning, it shifts and changes. Dr. Dowdye's *Effectivity* is not Galilean Relativity. Galilean Relativity relies on an invariant framework for light. Thus the title of the book to specify the rebirth of classical physics under a variable model of light. It is something new that should not have the name relativity associated with it at all.

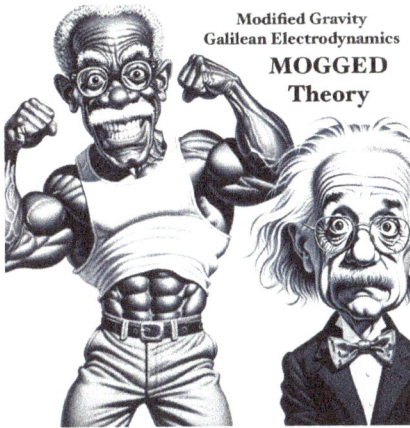

Fig. 1.4

This is not MOND Theory.
(Modified Newtonian Dynamics)

Dr. Dowdye's *Effectivity* under Galilean variance is its own revolutionary emissions theory. And it is indeed a modified understanding of Newtonian mechanics.

So, maybe call it:
*Modified Gravity Galilean Electrodynamics.*
(MOGGED)

The word "Relativity" will eventually be labeled a bane of the scientific community and have negative connotations in the future. Any framework based on an invariant model of light isn't based at all.

I aim to deliver what I call "*22nd Century Science*" which means getting rid of erroneous associations sooner than later. That abides by the logic of the whole "law of attraction" thing, right? Perhaps we need to act like we're already in the future by divorcing the outdated ball n' chain we call Relativity.

Fig. 1.5

In contrast to the Greek symbol Tau used by Effectivity, relativistic physics uses the standard lowercase letter t to represent the coordinate of time. That is the time according to a particular reference frame. Meaning the subjective experience of reality itself is claimed to be different for an observer moving toward or away from something he's looking at; and, whether the observer is accelerating or not towards or away in comparison to what he's looking at.

In classical physics $\tau_{tr}$ is used to describe a time interval that is invariant under Galilean transformations. Like the time between events as measured in the rest frame of an object. This goes back to the concept of absolute time from a Newtonian perspective, making for a rebirth of classical physics.

However, in relativity, time intervals can vary between subjective observers due to what relativists think is a literal warping or distorting of time/time dilation, and thus t is not invariant under Lorentz transformations. Time is variable and light is invariant under Lorentz transformations. Time is dependent upon the relative velocities between observers, distance, gravity, size, and the events being witnessed. Here are a few differences/ brief comparisons between the two models.

Fig. 1.6

Relativity and Lorentz invariance says:

$$c' = c$$

(The velocity of light is the same from all frames of reference)

but Classical Physics/Galilean variance says:

$$c' \neq c$$

(The velocity of light is NOT the same from all frames of reference. It's an illusion)

**I will discuss the apparent illusion of this scenario in great detail.**

Galilean variance defines $c'$ in this context as the following:

$$c' = c \pm v$$

(The velocity of light in one frame of reference is dependent upon the velocity of the light source relative to an observer in another frame of reference, whether that light source is approaching or receding away from that observer)

Albert Einstein's Relativity says:

$$E = mc^2$$

(Energy and mass are universally equivalent and literally interchangeable under All conditions.)

but Classical Physics/ Dr. Dowdye's Effectivity says:

$$E = \Delta mc^2 = m_o c^2$$

E equals the CHANGE of m c squared.

Energy changes in a system are the result from changes in mass. $m_o$ represents the original mass. Mass and energy do not literally interchange. There is an equivalence, not an interchange.)

Both Relativity & Classical Physics/Effectivity says:

$$\delta\theta = \frac{4GM}{Rc^2}$$

(But relativity says its gravity bending light/ the curvature of Space-Time... and classical physics says it's refraction ONLY at 1 solar radius.)

Relativity says the mass of a particle literally increases with acceleration:

$$m = \frac{m_0}{\sqrt{1 - \frac{v^2}{c^2}}}$$

But Classical Physics/ Effectivity says it's an illusion of surrounding potentials in a closed particle accelerator chamber. Not the mass itself. What is mistaken as a literal increase in mass is an artifact of the system's conditions... not a property inherent to the particle itself:

$$m_{eff} = \frac{m_0}{\sqrt{1 - \frac{v^2}{c^2}}}$$

So, rather than potentially having a knee-jerk reaction when seeing someone challenge Einstein, saying, "It must be wrong because it disagrees with relativity"... try saying, "How do Dr. Edward Dowdye's reformulated Galilean Transformations and Effectivity interpret the same scenario differently than relativity?"

The real question to ask is, "Do the equations accurately predict and match what is observed?" And the answer is a resounding, "Yes, they do!"

I will even go so far as to present how Galilean variance matches more so with experiments and observation than Relativity.

# Toothpick Bridge Analogy

Einstein built a bridge to the solution to the Perihelion of Mercury, PSR1913+16, and other interpretations to solve complex physics problems around the turn of the century.

Relativity and the approach of Lorentz invariance was the ONLY accepted method for over 100 years. Until Dr. Edward's reformulations first presented in 1991 in his book *The Extinction Shift Principle*. Although clever... relativity is not the right approach. (an artifice) And the bridge now has too many holes, collapsing under its own weight while the scientific community continues to walk in lockstep across it. All while the relativists cherry-pick aspects of Galilean invariance and Newtonian mechanics to use as glue. But there is now an airport on either side of the canyons. (Galilean variance).
The invariant toothpick bridge works... until it doesn't...

Fig. 1.7

Who was Dr. Edward Dowdye and what were his qualifications?

Dr. Edward Henry Dowdye Jr. was born in Washington DC December 13, 1943, and passed away December 31, 2020. He was the founder of *Pure Classical Physics Research* and *The Extinction Shift Principle*. Dr. Edward had a plethora of credentials.

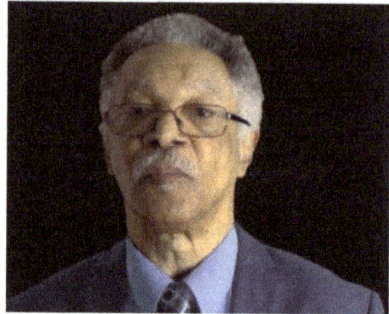

Fig. 1.8

Here is a list of some of his background and scholarly achievements:

BSc Mathematics - Hampton University

BSc Electronics Technology - Hampton University

BSc Physics - Hampton University

Diplom-Atomphysik (Atomic Physics) - Physikalisches Institut der Universität Heidelberg

A focus of Nuclear Magnetic Resonance on free ions/atoms

in plasma confinement chambers - Heidelberg & German Academic Exchange

Astronomy & Astrophysics - Universität Heidelberg

Ph.D. in LASER Spectroscopy Physics - Howard University

Electrical Engineering (Retired) - NASA Goddard, CSTEA

Taught Physics and Mathematics at Lincoln, Cheyney,

Marymount, Southeastern and Howard Universities as

well as instructing Atomic Physics students in Germany.

Fluent in English, German, Spanish and French

Member of the American Physical Society, SPIE and

other internationally recognized scientific organizations.

Dr. Dowdye has 16 Peer Reviewed Papers and over 40

Citations/ References/ Publications to corroborate.

(Listed at the end of the book)

Edward called his emissions theory, "*Effectivity*" and said it operates using classical physics. I am specifying that his *Extinction Shift Principle/Effectivity* operates under the framework of Galilean variance. Since both Galilean invariance and Galilean variance are both qualify as "classical physics." Which is why people's brains revert to autopilot when they hear the phrase "classical physics" or the word "Galilean". People are programmed to automatically think the words "Galilean" and "classical" and "Newtonian" and "Aether" and "Euclidean geometry" are one in the same. There are major differentiations to make here.

First was Newtonian mechanics within 3D Euclidean space. That is true classical physics. Then came the addition of aether and the application of Galilean invariant transformations within 3D Euclidean space. Still classical physics, but it mutated a bit. Then along came Einstein who screwed it all up with the invention of a 4th dimension and non-Euclidean geometry. (If it aint Euclidean... then it aint geometry...) Einstein would have been forbidden to enter Plato's school as per the plaque at the entrance stated.
"Let no one ignorant of geometry enter." --Plato

Then came along Dr. Edward Dowdye and his reformulation of classical physics. Returning back to 3D Euclidean space. Before the inappropriate coupling of space with time and the invention of the 4th dimension.

A brief history of physics: Newton/ classical physics/ 3D Euclidean ==> Aether/ Galilean invariance/ classical physics/ 3D Euclidean ==> Relativity/ Lorentz invariance/4D non-Euclidean ==> Effectivity/ Galilean variance/ classical physics/ 3D Euclidean

- Pinpointing flaws and correct aspects of physics throughout the eras.
- Addressing the history of where the scientific community went wrong in each era.
- Offering refinements to the similarities and tossing out the contradictions.
- Identifying the perpetrators who have forsaken nature's laws to maintain their mockery of real science.
- Why we should abandon everything from 1905 to 1990 and adopt everything from 1991 to the present. Returning to classical physics but under the mathematical framework of Galilean variance.

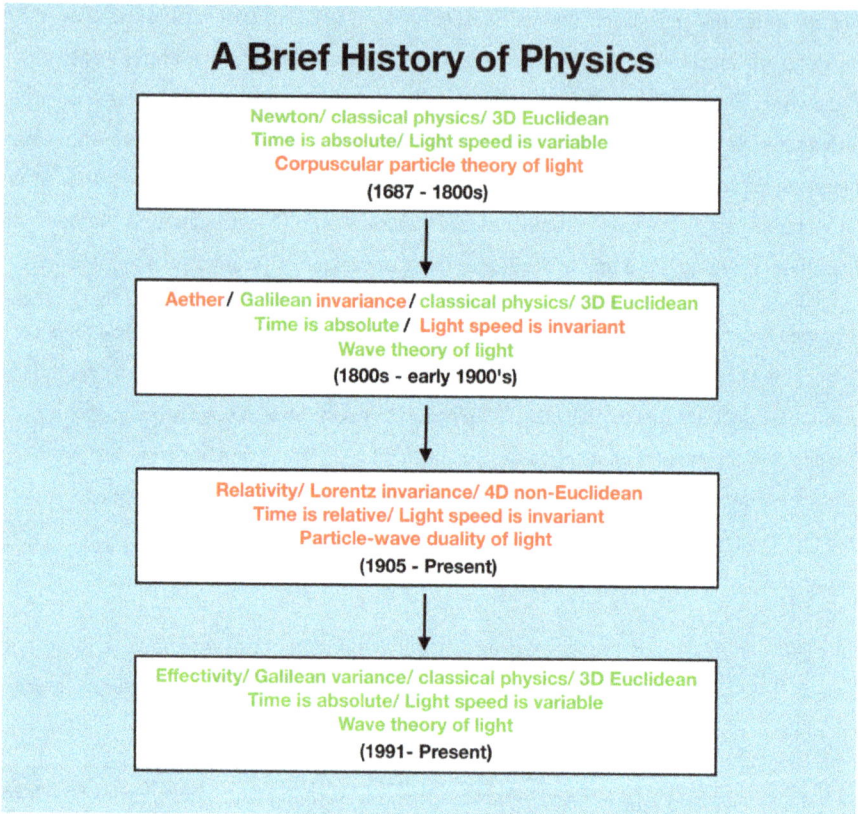

## A Brief History of Physics

Newton/ classical physics/ 3D Euclidean
Time is absolute/ Light speed is variable
Corpuscular particle theory of light
(1687 - 1800s)

Aether/ Galilean invariance/ classical physics/ 3D Euclidean
Time is absolute / Light speed is invariant
Wave theory of light
(1800s - early 1900's)

Relativity/ Lorentz invariance/ 4D non-Euclidean
Time is relative/ Light speed is invariant
Particle-wave duality of light
(1905 - Present)

Effectivity/ Galilean variance/ classical physics/ 3D Euclidean
Time is absolute/ Light speed is variable
Wave theory of light
(1991- Present)

Fig. 1.9

Why should we question Relativity and the use of Lorentz invariant transformations?
It is the scientific community's duty to scrutinize a model when there are volumes of new mathematical substantiations and justification, observational and experimental evidence, and a simpler method of yielding the same predictions and conclusions. William Occam would be proud.

The importance of Dowdye's Emission's Theory cannot be understated and must be taken into serious consideration. His unparalleled qualifications, scientific contributions and international recognition have earned the right to be heard.

There is so much politics in science now, it has created an idiocracy of doublethink where people now accept ideas that are mutually exclusive to reality. There is a tradition of nepotism where people hire their own who agree and carry on the invariant model of light. They keep the world filtering *all* interpretations through relativistic procedures and a Lorentz invariant model for light. The scientific hierarchy became a dogmatic echo chamber.

The people first encountering Dr. Edward Dowdye's Emissions Theory need to be mindful that they must cast aside ALL relativistic interpretations of space-time deformations, literal dilation of time or mass increase... and *any* extraneous mathematical requirements. All conjecture built upon the framework of an invariant model of light is erroneous.

This book will be reiterating and articulating interpretations through the lens of Galilean variance and the assumption that the velocity of light does indeed shift. Explanations for why will be given in detail. As Dr. Edward Dowdye wrote on his website ExtinctionShift.com in the FAQ section, the *Extinction Shift Principle* is "returning to the framework of Euclidean Space Geometry, essentially where we were before Poincaré, before Einstein and before the year 1913."

There has not been a more beneficial reformulation of science, math, and physics like this since Isaac Newton, Robert Hooke, Christiaan Huygens, or Nicolas Fatio. Even the work of Wolfgang Pauli and Walther Ritz seem to fall short in comparison to Dr. Edward Dowdye's approach. Since those ballistic theories and past emissions theories assume a literal particle-wave duality and that the speed of light stays the same no matter what. It does not. It's an illusion. Both light and gravity propagate at the rate of c, but relative to the source from which they are emitted.

For us to discuss gravity in detail though, we must first properly understand Light. Relativity does not adequately describe nature's processes, the propagation of light or the mechanism of gravity. Lorentz invariance has now created more problems than it set out to initially solve. All because Einstein refused to acknowledge that the velocity of light shifts. Relativity simply repackaged the 3D luminiferous aether as a 4D space-time.

# Would YOU Buy a Used Theory From This Man?!

Fig. 1.10

Aether was proposed under Galilean invariance to justify the speed of light being the same no matter what... to explain how light itself propagates and how gravity influences over great distances. So, they thought a 3D mass displaces a 3D "luminiferous aether medium". But displacement is simultaneous and instant. Like sitting in a bathtub and the water level of the total volume rises in proportion, and that requires a condensed matter surface to push off of. And if you want curvature of the medium, it's going to be spinning and pushing against the walls of its container through convection.

When Michaelson-Morley's experiments produced nullified results, Lorentz invariance was invented. The concept of a deforming space-time was invented as an artifice to justify the speed of light being the same, no matter what. The error is in the assumption that light speed is invariant and requires ANY medium at all!

Around 1911, the 3D Aether medium was replaced with a 4D curving space-time. Instead of Galilean invariance using aether... Einstein invented Lorentz invariance using the curvature of space-time. And relativists (they) define gravity itself as the deformation of a 4D medium in the presence of a 3D mass. But there is no such medium, no displacement and gravity is a force and an emission. Both relativity & aether are based on invariant models of light that are destined to oblivion.

Dr. Edward Dowdye's Extinction Shift Principle gives justification to drop both Relativity and aether and ANY invariant model for light requiring *any* medium to propagate light or generate gravity. The 4th dimension is an invention to try and save face for the dismantled 3D aether medium. Space-time does not exist and neither does aether. Sorry Einstein. Sorry Tesla. The same light does not remain the same velocity ... and yet distort within the same frame of reference because of a deforming medium that can never be tested in a lab. How convenient.

Nikola Tesla's quotes on Relativity in an article with the New York Times dated July 11th, 1935: "On a body as large as the sun, it would be impossible to project a disturbance of this kind [e.g., radio broadcasts] to any considerable distance except along the surface. It might be inferred that I am alluding to the curvature of space supposed to exist according to the teachings of relativity, but nothing could be further from my mind. I hold that space cannot be curved, for the simple reason that it can have no properties. It might as well be said that God has properties. He has not, but only attributes and these are of our own making. Of properties we can only speak when dealing with matter filling the space. To say that in the presence of large bodies space becomes curved, is equivalent to stating that something can act upon nothing. I for one, refuse to subscribe to such a view." -- Nikola Tesla

Around 1916, Lorentz invariance replaced the outdated aether theory under Galilean invariance. Einstein's theory of General Relativity transmuted the luminiferous aether into a 4D curved space-time. This neo-aether is no longer viewed as a medium in 3D Euclidean space, but now in a fourth-dimensional and non-Euclidean deforming space-time." AS IF a

3D medium. Referring back to the toothpick bridge analogy, relativity cherrypicked aspects of 3D physics to hold its own 4D artifice together. It was the invention of the 4th dimension that completely disgusted Tesla and he brutally criticized Einstein in the 1930's because of it. In that same interview he gave to the New York Times on July 11th, 1935, Tesla is quoted as saying , "The Theory of Relativity is just a mass of error and deceptive ideas violently opposed to the teachings of great men of science of the past and even to common sense. The theory wraps all these errors and fallacies and clothes them in magnificent mathematical garb which fascinates, dazzles and makes people blind to the underlying error. The theory is like a beggar clothed in purple whom ignorant people take for a king. Its exponents are very brilliant men, but they are meta-physicists rather than scientists."

In Tesla's "*Dynamic Theory of Gravity*", he stated, "Supposing that the bodies act upon the surrounding space causing curving of the same, it appears to my simple mind that the curved spaces must react on the bodies, and producing the opposite effects, straightening out the curves. Since action and reaction are coexistent, it follows that the supposed curvature of space is entirely impossible. But even if it existed it would not explain the motions of the bodies as observed. Only the existence of a field of force can account for the motions of the bodies as observed, and its assumption dispenses with space curvature. All literature on this subject is futile and destined to oblivion. So are all attempts to explain the workings of the universe without recognizing the existence of the aether and the indispensable function it plays in the phenomena. My second discovery was of a physical truth of the greatest importance. As I have searched the entire scientific records in more than a half dozen languages for a long time without finding the least anticipation, I consider myself the original discoverer of this truth, which can be expressed by the statement: There is no energy in matter other than that received from the environment."
-- Nikola Tesla, 1931

When using an invariant mathematical model, whether it be Galilean invariance or Lorentz invariance, to deny the existence of SOME medium like aether, that interpretation built upon the denial of the aether is destined to oblivion. The denial of aether is a violation under an invariant model of light. But Nikola Tesla was not aware of a variant model of light which justifies the total abandonment of any requirement for an aether or any medium. But you can't violate something that doesn't apply. Under Galilean variance, the workings of the universe do not rely on any aether, whether it be static or dynamic. Galilean variance doesn't rely on any medium at all, or the existence of a fourth dimension, no need for the existence of space-time or anything like that. Aether and space-time is an excuse to interpret light as being the same velocity in all frames of reference no matter what. So Tesla is correct when saying the aether is indispensable to the function of the universe.... under the assumption that the velocity of light remains the same. But it does not. It's an illusion. Therefore, Tesla's assumption, that there MUST be an aether to explain the workings of the universe, is erroneous. The assumption that a 4th dimensional medium is erroneous. A great resource about the delusion of the 4th dimension is a video by Matt Presti called Dispelling Dimensional Madness.

https://www.youtube.com/watch?v=20SW9smqqfg

# Misconception of Aether

I used to believe in aether. But now I think... if you can define it, then it isn't aether.
I think aether is like the word quantum or magic. It's an esoteric and convenient copout to
avoid talking about particle physics and dynamics on scales smaller than what Max
Planck's models would allow. And describes the behavior of charged particles below
"ground states" that "shouldn't be possible." The existence of an aether is a must and
requirement under Galilean in variance. It's a consequence of the model. A shift in
frequencies is observed between clocks at different altitudes, therefore... aether.
Or the consequence of the model of Lorentz invariance is that the shift in frequencies
observed between clocks is therefore... space-time.

Anything not understood is automatically labeled quantum or aether or magic, etc. And if
there's enough mass present, the same mysteries are attributed to space-time or dark
matter. I think there are fine and ultra-fine particles of matter like what Nicolas Fatio and
Georges LeSage speculated. I think there are gases that pervade space that are not
recognized. Like what Walter Bowman Russell presented on his periodic chart from 1926,
in his book, *A New Concept of the Universe*. He showed 24 elements preceding hydrogen.
Unacknowledged rarified gases could have been considered the aether.

Thousands of years ago, ancient Greeks thought plasma was aether. But as understanding
grew and scientific tools evolved, our terminology and understanding also advanced. What
used to be "aether" evolved into a now definable and a definitive variable. I view aether as
being undefinable. When you give a definition to something, it shifts the goalposts of what
aether is.

It's debatable that even Nikola Tesla and Charles Steinmetz viewed an electrostatic field/
electrostatic potential as being "the aether." So, if there are particles not recognized... and
gases not recognized... and an all-pervasive plasma throughout the galaxy with static
electric fields that extends way past the boundary of the mass generating that influence...

and given the semantics of all those variables... then those are real factors that we can talk about. Therefore, the more we learn, the less "aether" plays any role. It's not needed under this specific framework of Galilean variance. In this model, we simply acknowledge the velocity of light shifts. That's it. No need for aether, space-time or any model relying on a medium as an excuse for light to remain the same in all frames of reference or to dismiss a frequency shift as anything other than the speed of light itself shifting. Aether doesn't play any role under Galilean variance. Aether is specific to Galilean **INVARIANCE**. And neither does dark matter play any role in Galilean **variance** or a **variable** model of light. Dark matter, dark energy and space-time is specific to Lorentz **INVARIANCE**. And so is pretty much every other "modern" concept based upon an erroneous foundation of an invariant model of light.

Aether was proposed to try and explain how light propagates and how gravity influences over great distances. So they thought a 3D mass displaces a "luminiferous aether".
Like how you sit in a bathtub and the water level rises in proportion. They viewed gravity as being an instantaneous action as a result from a 3D mass displacing an aether medium. But there is no such medium and gravity is not instant. Not to mention that the water level rising simultaneously requires the walls of a bathtub for the water to push against. That forces the total volume of water to displace and rise up and out of the condensed matter tub. But in relativity they say space-time itself acts like the surface of water or the surface of a mattress. And in the presence of a 3D mass, space-time displaces and curves. Without need for any container or mass to push against to cause curvature or displacement of said space-time. Too many contradictions to list, I digress.

In 1911, that 3D Aether medium was replaced with a 4D curving space-time. Instead of light itself as being the deformation of a rippling or buckling aether, or light propagating through "the aether," it's said light propagates AROUND the curvature of space-time. And relativists say the curvature of space-time IS gravity.

But none of that is real. Light only refracts in space from the plasma around the surface of a star. Doesn't progressively "bend" away from a star. And classical physics doesn't require

ANY media for light or gravity to propagate. That's part of the error. Which means the scientific community's search for an explanation of a non-existent medium are ultimately futile. Oh the irony Mr. Tesla. All models based on a 4D curving space-time AND an aether medium or any previous emissions theories are destined to oblivion. As Galilean transformations under an invariant interpretation are also destined to oblivion... in addition to QED which was built upon Lorentz invariant transformations from the original assumption of the aether's existence.

"Galilean relativity" relied on an aether medium to propagate light. But the math for original Galilean relativity stated the velocity of light is the same to everyone in all frames of reference no matter what. And that an aether medium was responsible for the apparent constant in all frames. They couldn't find evidence of any aether medium (nor any medium at all) from the Michaelson-Morley experiments, Hammar experiments or Dayton Miller experiments. That led to the abandonment of aether theory. And further, that led to the invention of relativity under Lorentz transformations. It also led to the invention of MOND theory, tired light theory and a lot more.

The velocity of light is the same to everyone *no matter what* in relativity and the wave is still invariant. But now the cause is a 4th Dimensional space-time medium instead of a 3D aether medium. Treating it just like the aether but calling it a new name with more devious equations to substantiate it. Regardless, for the entirety of aether theory, relativity theory and quantum theories, string theory, etc. ALL of them rely on the INVARIANCE of a wave and that the speed of light is the same and unchanging to *all* observers *no matter what.*

Finally, we can return to classical physics once again, but under the framework of Galilean VARIANCE. Correcting the initial mistakes which caused relativity to be invented in the first place. And everything built upon Lorentz transformations collapses. All of it. Don't be sad... rejoice! This means more people will be toasty warm in the winter because there are so many diplomas and PhD certificates to burn, we could turn Antarctica into a paradise. The entire foundation and reason for the existence of relativity and QED has been retroactively wiped out and dismantled by Dr. Dowdye's introduction of

GALILEAN VARIANCE. There's hundreds of years of discoveries to be made again by reinterpreting ALL past empirical data and evidence. The entire concept of empirical evidence itself is turned on its head, since evidence and proof are two different things and experiments never prove a theory. The same empirical evidence can be reinterpreted under a mutually exclusive mathematical framework which can disprove and dismantle a previously accepted theory. The term "empirical evidence" is wrongfully used to suggest "infallible."

Relativity and QED derived interpretations AS IF there is a 4th dimension.
AS IF mass increases with acceleration...
AS IF time literally dilates...
AS IF gravity bends light....
AS IF there is a particle-wave duality...
AS IF the wave is invariant...
AS IF there was a light speed limit...
AS IF dark matter/energy was real...
AS IF black holes exist...
Etc., etc., and so forth.

But it's all based on an artifice assuming an invariance of the wave, whether it be from aether or space-time curvature.
But it's none of that. Because the wave IS indeed variant. No medium needed at all.

AS IF ≠ Literal.

Fig. 1.11

— CHAPTER 2 —

# Reflection Doesn't Exist

Reflection technically doesn't exist because the same light does not ricochet. It's never the same light. The following formula is an example of an equation derived by Dr. Edward Dowdye, presenting an alternative to Doppler Shift and the Invariance of the Wave Equation under Lorentz invariance In the following version of the wave equation under reformulated classical physics, the velocity of the wave is variable.

All the light you ever observe is the re-emitted light from the electrons making up yourself. Electrons absorb and emit all light and electromagnetic radiation.
All matter is made of atoms. All atoms contain electrons.
Electrons absorb all incoming primary packets and re-emit a secondary packet from their own frame of reference.

It is not the same light coming in through your window into your eyes from the same frame of reference. No primary wavepacket can ever distort. A brand-new secondary packet is re-emitted instead, serving the illusion of distortion. When a packet of light shifts frames of reference, it changes frequency. And that change in frequency IS a change in the velocity of light.

Here is an alternative to the wave equation under Galilean variance. Dr. Dowdye's procedure yields the same results without needing relativity's extraneous corrections... And without the illusions. The wheat without the chaff.

$$\frac{\partial^2 \Phi}{\partial x^2} + \frac{\partial^2 \Phi}{\partial y^2} + \frac{\partial^2 \Phi}{\partial z^2} - \frac{1}{c^2}\frac{\partial^2 \Phi}{\partial t^2} = 0$$

The IMAGINARY OBSERVER gets:

$$\Phi = \Phi_0 \sin 2\pi \left( vt + \frac{1}{\lambda} x \right)$$

is a solution of the wave equation of the PRIMARY wavepacket at velocity C relative to a Source S, where $v\lambda = c$, (because it's coming from the source itself) but at velocity $c'$ (from another frame of reference) $\neq c$ relative to the rest frame.

The ACTUAL OBSERVER gets:

$$\Phi' = \Phi'_0 \sin 2\pi \left( v't' + \frac{1}{\lambda}' x' \right)$$

That works to be a solution of the wave equation of the SECONDARY re-emitted wave at velocity c relative to the Source S' in another frame of reference.

Important to note: There is NO Time Dilation in Euclidean Space under the electrodynamics of Galilean variance. There is an equivalent "Transverse Relative Tim e Shift" between CLOCKS but not a dilation of reality itself. The difference in frequencies between atomic clocks IS a literal change in the velocity of light. But because relativity forbids the velocity of light from ever-changing, people wrongfully assume the face of a clock represents reality itself. The data transmitted between satellite and ground does not dilate at all though.

Thus, $$t' = t$$

(Time **IS** the same from all frames of reference, not the velocity of light)

Both the IMAGINARY and ACTUAL OBSERVER would find that the velocity of the wave being observed is always:

$$v'\lambda' = [v(1 \pm cv)][\lambda(1 \pm cv)^{-1}] = v\lambda = c$$

relative to the observer's CLOSEST primary source.

(Which is ultimately the electrons making up itself... from its own frame of reference)

Differentiating the equation for Φ twice after t and x, the IMAGINARY OBSERVER derives:

$$\frac{\partial^2 \Phi}{\partial t^2} = -\Phi(2\pi)^2 v^2 = v^2 \lambda^2 \frac{\partial^2 \Phi}{\partial x^2}$$

And

$$\frac{\partial^2 \Phi}{\partial x^2} + \frac{\partial^2 \Phi}{\partial y^2} + \frac{\partial^2 \Phi}{\partial z^2} - \frac{1}{v^2 \lambda^2} \frac{\partial^2 \Phi}{\partial t^2} = 0$$

The ACTUAL OBSERVER derives:

$$\frac{\partial^2 \Phi'}{\partial x'^2} + \frac{\partial^2 \Phi'}{\partial y'^2} + \frac{\partial^2 \Phi'}{\partial z'^2} - \frac{1}{v'^2 \lambda'^2} \frac{\partial^2 \Phi'}{\partial t'^2} = 0$$

Meaning,
$$t' = t$$

(Time **IS** the same from **ALL** frames of reference)

$$c' \neq c$$

(The velocity of light is **NOT** the same from all frames of reference)

$$c' = c \pm v$$

(The velocity of light in one frame of reference **IS dependent upon the velocity of the light source** relative to an observer in another frame of reference, whether that light source is approaching or receding away from that observer)

**Time** is the same from **ALL** frames of reference, **NOT** the velocity of light!

• Both mathematical frameworks of Lorentz transformations and Galilean transformations yield the same predictions and solutions now thanks for Dr. Edward Dowdye's reformulations in 1991. But both frameworks are mutually exclusive and interpret the entirety of how the end results of a scenario arise completely differently.

• The illusion of light seemingly being invariant is that no matter what... ALL observers find the velocity of the wave to *always* be C.

(Only apparent relative to each observer's point of view from THEIR most primary source.

That "most primary source" ultimately ends up being the electrons making up ourselves.)

$$v'\lambda' = \left[v\left(1\pm\frac{v}{c}\right)\right]\left[\lambda\left(1\pm\frac{v}{c}\right)^{-1}\right] = v\lambda = c \;\therefore\; v'\lambda' = v\lambda = c$$

(The velocity of light is measured as C in all frames. But it's not the same light that distorts or ricochets off a mirror and continues-on into your eye. It is a brand-new light/re-emission at each point of interference. Every point of interference generates re-emissions which then travel relative to that new frame of reference.)

As Albert Einstein himself wrote in a letter to Erwin Finley-Freundlich in August of 1913:
*"Wenn die Lichtgeschwindigkeit auch nur ein bißchen von der Geschwindigkeit der Lichtquelle abhängig ist, dann ist meine ganze Relativitätstheorie und Gravitationstheorie falsch."*

**"If the velocity of light is even a little bit dependent upon the velocity of the light source, then my ENTIRE theory of Relativity and theory of Gravitation is FALSE." -- Einstein**

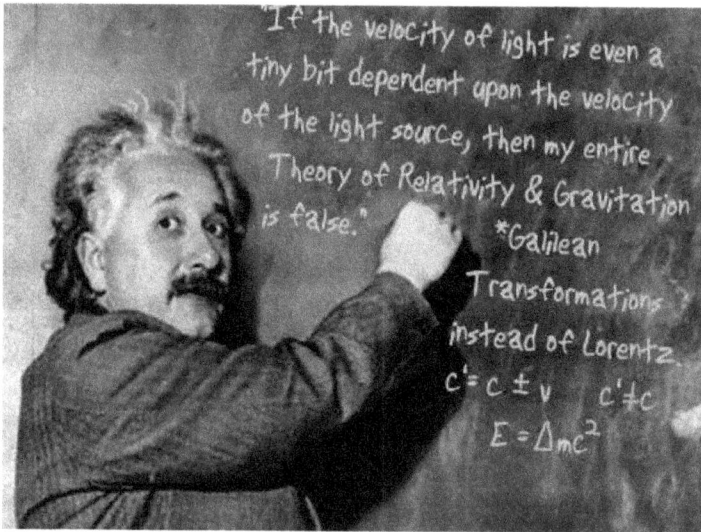

Fig. 2.1

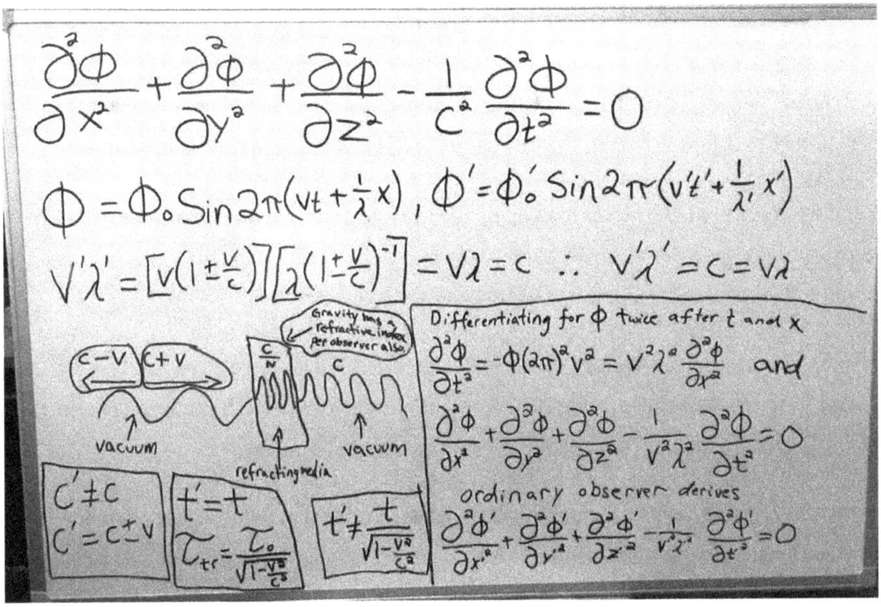

$$\frac{\partial^2 \phi}{\partial x^2} + \frac{\partial^2 \phi}{\partial y^2} + \frac{\partial^2 \phi}{\partial z^2} - \frac{1}{c^2}\frac{\partial^2 \phi}{\partial t^2} = 0$$

$$\phi = \phi_0 \sin 2\pi\left(vt + \frac{1}{\lambda}x\right), \quad \phi' = \phi_0' \sin 2\pi\left(v't' + \frac{1}{\lambda'}x\right)$$

$$v'\lambda' = \left[v\left(1\pm\frac{v}{c}\right)\right]\left[\lambda\left(1\pm\frac{v}{c}\right)^{-1}\right] = v\lambda = c \therefore v'\lambda' = c = v\lambda$$

Gravity bends refractive index per observer also

$c-v$ $c+v$ → vacuum, $\frac{c}{n}$ $c$ refracting media vacuum

$c \neq c$, $c' = c \pm v$

$t' = t$, $t_{tr} = \frac{\tau_0}{\sqrt{1-\frac{v^2}{c^2}}}$

$t \neq t$, $\frac{t}{\sqrt{1-\frac{v^2}{c^2}}}$

Differentiating for $\phi$ twice after $t$ and $x$

$$\frac{\partial^2 \phi}{\partial t^2} = -\phi(2\pi)^2 v^2 = v^2\lambda^2 \frac{\partial^2\phi}{\partial x^2} \quad \text{and}$$

$$\frac{\partial^2\phi}{\partial x^2} + \frac{\partial^2\phi}{\partial y^2} + \frac{\partial^2\phi}{\partial z^2} - \frac{1}{v^2\lambda^2}\frac{\partial^2\phi}{\partial t^2} = 0$$

ordinary observer derives

$$\frac{\partial^2\phi'}{\partial x^2} + \frac{\partial^2\phi'}{\partial y^2} + \frac{\partial^2\phi'}{\partial z^2} - \frac{1}{v^2\lambda^2}\frac{\partial^2\phi'}{\partial t^2} = 0$$

Fig. 2.2

If we trace the path of light from a bulb or a source… we would see the light emitted out like a shockwave at the velocity of C. Each point on the wavefront can be traced back to the source in a rectilinear, or straight-lined path. The light is projected out in a straight line, from any given point of the source… forming a total expanding sphere of electromagnetic radiation. **Any** time that straight lined path of light is blocked or interfered with… that light will be absorbed by the electrons making up the blockage. That even means your eye.

It is not the same light coming from the sun, through your window and into your eye. The sun produces a brand-new light at a constant. That light propagates out at the velocity of C and takes about 8.33 minutes to reach Earth. The electrons making up Earth's atmosphere absorbs that incoming constant energy. But those electrons can only absorb so much before they must re-emit again. And they do so in a wave-like manner, producing re-emissions as a frequency of interference between the sun and themselves.

Those re-emissions propagate through Earth's atmosphere... generated by the electrons making up Earth's atmosphere itself, and then that light is absorbed by the electrons making up the glass of your window. The electrons making up your window will re-emit yet another brand-new tertiary (third) packet of light which then travels to your eye.

But if we zoom in on the atoms making up the eye itself... we would note the electrons making up the observer absorbs that tertiary packet... and simultaneously reproduces a brand-new quaternary (fourth) as an equal and opposite reaction.

I liken the process to an analogy of two balloons connected end to end. As one balloon deflates... the second balloon simultaneously inflates in proportion from the displacement of air. The air from one balloon fuels the other as the air diminishes.

As the energy of the primary emission diminishes to zero from being absorbed by the electrons making up an obstruction... a new secondary packet simultaneously crescendos in proportion as a new re-emission. That process happens at the rate of C... serving the *illusion* that the velocity of light is the same to all observers in all frames of reference.

There is a difference between the velocity of light constant ($\approx 299{,}792{,}458$ meters per sec/ 186,282 miles per second) versus the constancy of the velocity of light in ALL frames of reference. The velocity of light constant, C, is well established. The problem is the illusion of re-emissions from the electrons making up each observer and detector. The constancy of light SEEMS the same to all observers, but that's because the act of measurement itself introduces interference which produces re-emissions and secondary packets.

It is physically impossible for a sensor to measure the velocity of any primary wave packet or original light from a source. Relativity and Lorentz transformations view the same scenario as THE SAME LIGHT traveling out... getting interfered with... distorting, and then continuing-on within the same frame of reference as a distorted primary. But under the framework of Galilean transformations, a primary can never distort.

All primaries are absorbed... become extinct and a new light or packet of electromagnetic radiation is generated from that new point. The same light has not existed since some "Big Bang" and traveled around since the dawn of the universe, bouncing, distorting, and continuing-on within the same frame of reference.

That simply does not happen under the framework of Galilean variant transformations. Lorentz invariant transformations can calculate the scenario AS IF it's the same light... but as Dr. Edward wrote in that same FAQ section of this site, "... *pertaining to the interstellar and intergalactic media.... It is clearly seen that the primary emissions from the direct sources of emissions stand a ghost of a chance reaching any Earth based observers, as the primary emissions are immediately extinguished by the predominant unseen matter, the gases and "secondary sources" of the interstellar and intergalactic media."*

So, to put it succinctly... relativity says light is invariant, meaning, the speed of light is the same to *everyone* no matter what. Doesn't matter if someone is accelerating toward or away from a light source. Doesn't matter of the frequency shifts or stars changing position during a solar eclipse. Relativity says the speed of the light you are running into is the same speed no matter what, and everything we observe is attributed to the deformation of a 4D "fabric" of space-time. But the Extinction Shift Principle/ Effectivity under Galilean transformations say the invariance of light is only apparent... and an illusion, due to the wrongful assumption that the same primary light is continually being observed in all frames of reference no matter what.

Dr. Dowdye's approach under reformulated Galilean transformations correctly differentiates between the illusion of the re-emissions due to the electrons making up the observer himself verses the actual variable velocity of light between frames of reference.

Here's another quick example: All the light you ever see is the re-emitted light from the electrons making up yourself. It is never the same light coming from a flashlight into your eyes. It's never the same light bouncing off of a mirror. It's never the same light passing through a lens or refracting. It's always a brand-new light at every point of interference.

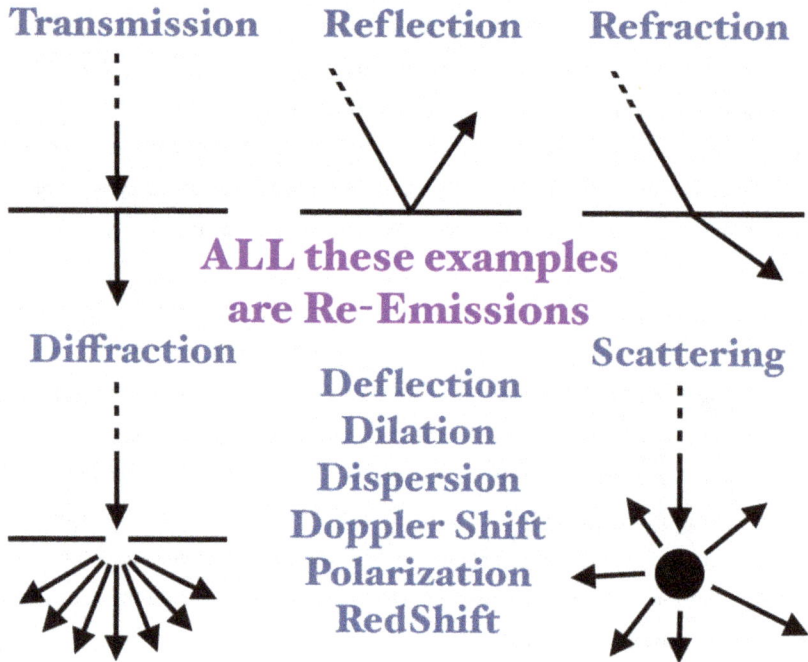

Fig. 2.3

It is never the same light propagating from a flashlight to your eye. Any obstructions/blockage/interference will result in the primary light being absorbed by the electrons making up the obstruction. A brand-new secondary light is re-emitted as an equal and opposite reaction. The re-emitted light does not always travel relative to the primary light.

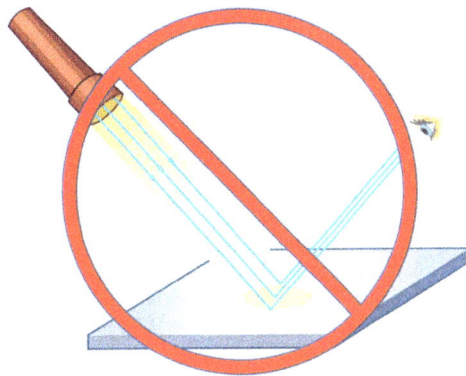

Fig. 2.4

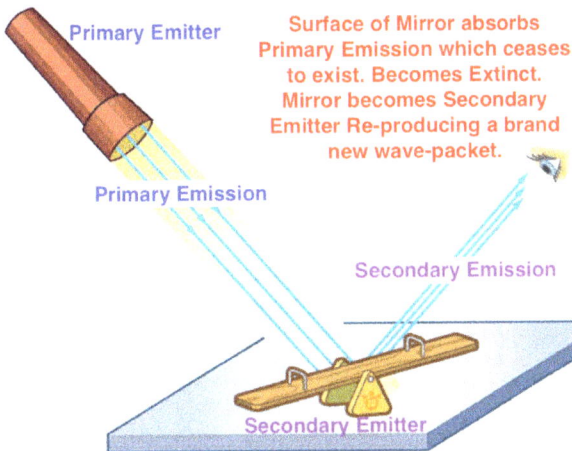

Primary Emitter

Surface of Mirror absorbs Primary Emission which ceases to exist. Becomes Extinct. Mirror becomes Secondary Emitter Re-producing a brand new wave-packet.

Primary Emission

Secondary Emission

Secondary Emitter

Fig. 2.5

The light coming out of the flashlight is propagating at the rate of **C**. Then, the electrons making up the surface of the mirror absorbs that incoming primary which diminishes to zero. And as an equal and opposite reaction, a secondary light is re-emitted from those same electrons and crescendos at the rate of **C**. Those electrons travel relative to the surface of the mirror, thereby producing a re-emission propagating relative to the mirror.

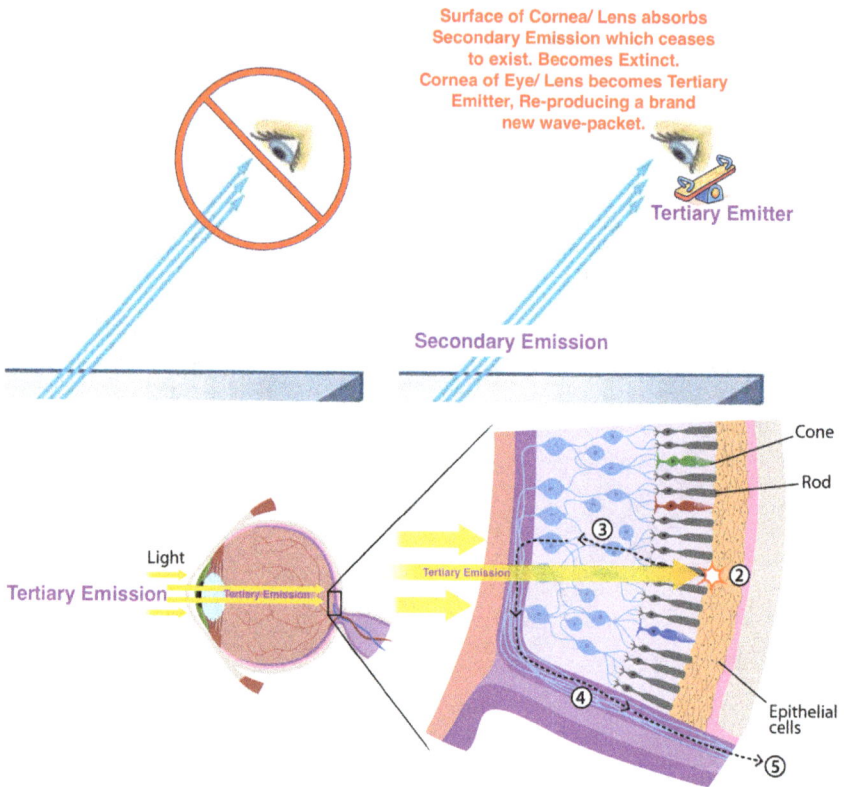

Fig. 2.6

Then the electrons making up your eye absorbs that secondary light and re-emits yet another third wave packet/tertiary light. That new tertiary/third light will be re-emitted by the electrons making up the rods and cones of your eye itself. And that final re-emission is what you detect and measure.

# Constant from Source
## vs
## Constancy in ALL Frames of Reference

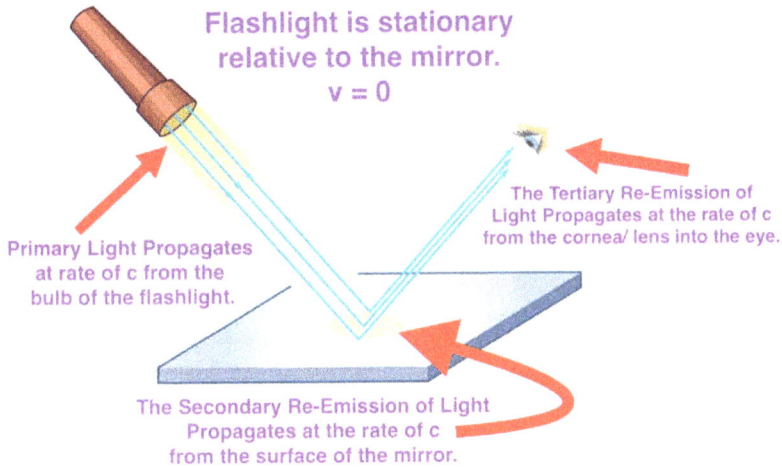

Flashlight is stationary
relative to the mirror.
v = 0

The Tertiary Re-Emission of
Light Propagates at the rate of c
from the cornea/ lens into the eye.

Primary Light Propagates
at rate of c from the
bulb of the flashlight.

The Secondary Re-Emission of Light
Propagates at the rate of c
from the surface of the mirror.

Fig. 2.7

If the flashlight is stationary relative to the mirror, the light from that flashlight will be in the same frame of reference as the mirror. The light from that flashlight propagates out at the rate of C. The electrons making up the mirror absorb that incoming primary light and re-emits a brand-new secondary light from its own frame of reference. If your eye is stationary relative to the mirror, then your eye is in the same frame of reference as the mirror *and* the flashlight in this scenario. The primary light propagates from the flashlight at C. The mirror absorbs and re-emits at the rate of C. The secondary light from the mirror propagates out at C. The electrons making up the eye absorb and re-emit a brand-new tertiary/third light at C.

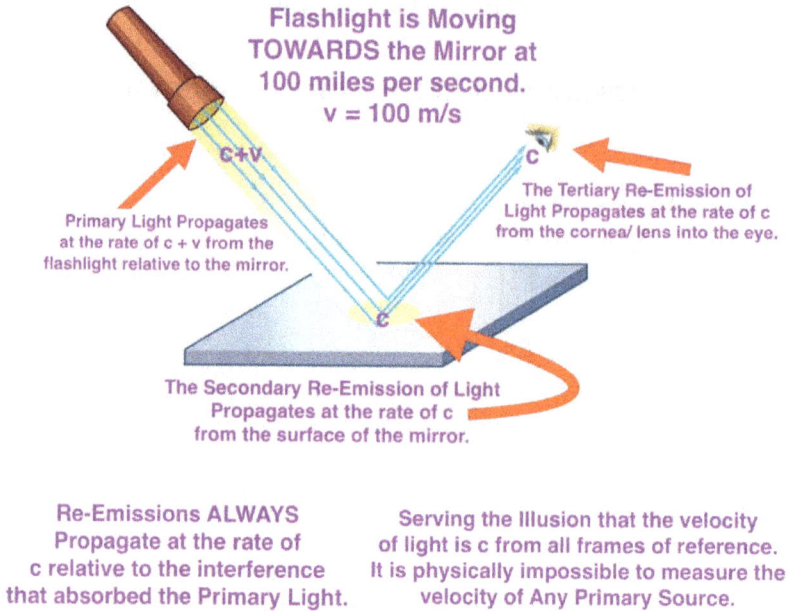

**Flashlight is Moving TOWARDS the Mirror at 100 miles per second.**
**v = 100 m/s**

c+v

Primary Light Propagates at the rate of c + v from the flashlight relative to the mirror.

The Tertiary Re-Emission of Light Propagates at the rate of c from the cornea/ lens into the eye.

c

The Secondary Re-Emission of Light Propagates at the rate of c from the surface of the mirror.

Re-Emissions ALWAYS Propagate at the rate of c relative to the interference that absorbed the Primary Light.

Serving the Illusion that the velocity of light is c from all frames of reference. It is physically impossible to measure the velocity of Any Primary Source.

Fig. 2.8

If the flashlight is moving towards the mirror, then the light coming from that particular flashlight is traveling C *plus* the velocity of the flashlight itself relative to the mirror it's approaching. But by the time the light actually hits the mirror, the electrons making up the mirror will absorb that light and re-emit a secondary from its own frame of reference. Then the electrons making up your eye will absorb that secondary and re-emit a third light. It is not a Doppler Shift since that model assumes the same approaching light is the same distorted light that reaches your eye. It's *Extinction Shifted.* A re-emission. A new light.

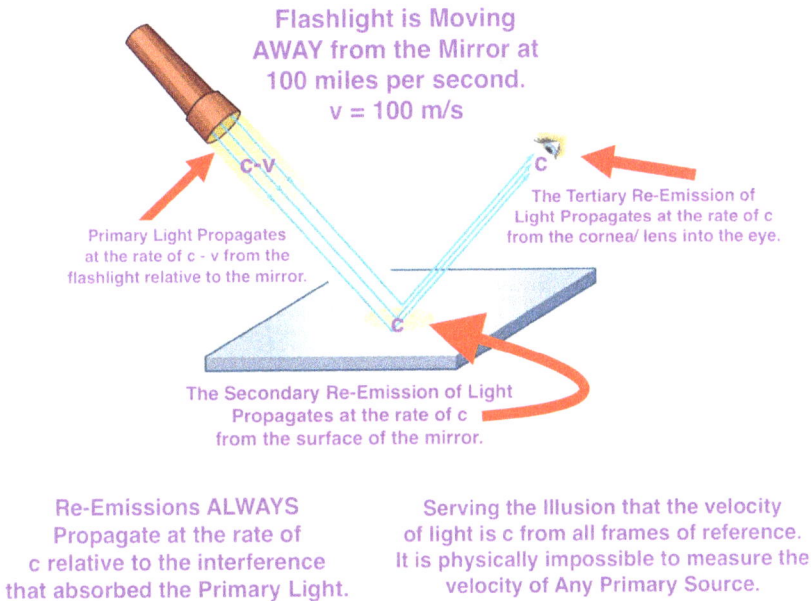

Flashlight is Moving
AWAY from the Mirror at
100 miles per second.
v = 100 m/s

c-v

Primary Light Propagates
at the rate of c - v from the
flashlight relative to the mirror.

c

The Tertiary Re-Emission of
Light Propagates at the rate of c
from the cornea/ lens into the eye.

c

The Secondary Re-Emission of Light
Propagates at the rate of c
from the surface of the mirror.

Re-Emissions ALWAYS
Propagate at the rate of
c relative to the interference
that absorbed the Primary Light.

Serving the Illusion that the velocity
of light is c from all frames of reference.
It is physically impossible to measure the
velocity of Any Primary Source.

Fig. 2.9

If the flashlight is moving away the mirror, then the light coming from that particular flashlight is traveling C *minus* the velocity of the flashlight itself relative to the mirror it's receding away from. By the time the light actually hits the mirror, the electrons making up the mirror will absorb that light and re-emit a secondary from its own frame of reference. Then the electrons making up your eye will absorb that secondary and re-emit a third light. Again, it is not a Doppler Shift since that model assumes the same approaching light is the same distorted light that reaches your eye. It's *Extinction Shifted.* A re-emission/ new light.

By the time *any* light, reaches your eye or a detector, it will be re-emitted as the frequency of interference between the source and the detector. And you will *never* be able to measure the velocity of that primary. But you *can* see a flash before one source or another if one of those sources is traveling toward you and the other is not. But whoever measures any incoming flashes will still measure them all as C, no matter what.

We can briefly compare to the relativistic interpretation under Lorentz invariance:

$$\frac{\partial^2}{\partial x^2} + \frac{\partial^2}{\partial y^2} + \frac{\partial^2}{\partial z^2} - \frac{1}{c^2}\frac{\partial^2}{\partial t^2} = \frac{\partial^2}{\partial x'^2} + \frac{\partial^2}{\partial y'^2} + \frac{\partial^2}{\partial z'^2} - \frac{1}{c^2}\frac{\partial^2}{\partial t'^2}$$

When really it should say, does NOT equal:

$$\frac{\partial^2}{\partial x^2} + \frac{\partial^2}{\partial y^2} + \frac{\partial^2}{\partial z^2} - \frac{1}{c^2}\frac{\partial^2}{\partial t^2} \neq \frac{\partial^2}{\partial x'^2} + \frac{\partial^2}{\partial y'^2} + \frac{\partial^2}{\partial z'^2} - \frac{1}{c^2}\frac{\partial^2}{\partial t'^2}$$

The velocity of light is always relative to the source from which it originates. As a brand-new light is continuously emitted from a source, it propagates with a velocity that is dependent on the source's velocity at the precise moment of emission, regardless of the source's acceleration. Therefore, $c' = c \pm v$. Consequently, this newly produced light maintains a consistent trajectory in a straight line/ rectilinear path relative to the source's motion at each point along the source's path.

To put simply, the light from a given source will move faster than C compared to you if that source is moving toward you. Like how the bullet from a gun will hit you with a higher velocity if someone is shooting at you while they are on a train heading towards you.

The light from a given source will move slower than C compared to you if that source is moving away from you. Like how a bullet from a gun will have less speed compared to you if you are traveling away from the bullet as it approaches you. The emission is still coming out of the gun or the light source at the same velocity... but relative to YOU it could be faster or slower if you're heading towards or away from the source its emission.

The illusion is that all observers and detectors are only capable of measuring the re-emitted light from the electrons making up themselves. So even though the wavepacket could be traveling $c \pm v$ relative to an observer... the electrons making up that observer will absorb that incoming primary... which will diminish to zero.

Those same electrons will crescendo a brand-new secondary packet as an equal and opposite reaction, and that new re-emitted secondary wavepacket will be propagating relative to the electrons making up that observer. That process happens at $c$. And since the electrons of an observer are traveling relative to that observer... it serves the illusion that ALL wavepackets are the same velocity and that the velocity of light is invariant in ALL frames of reference to ALL observers. It is not. It's an illusion dealing with the re-emission of light from the electrons making up ourselves.

All the light you ever see is ultimately the re-emitted light from the electrons making up yourself. All observers only assume they are measuring the primary packet because the velocity of light is $c$ to all who measure. But that is only apparent due to the re-emission from one's own electrons. Imagine you're encased in a block of colored glass. You can only measure the refractive index of the medium surrounding you. The velocity of light is propagating at the rate of $\frac{c}{n}$ within the medium. Outside of the medium... the source emitting the primary packet might be approaching you or receding away from you. You'd have no way of knowing that.

Therefore, the emissions coming from that approaching or receding source might be propagating at $c \pm v$ (depending on if the source is approaching toward or receding away from the medium encasing you.) So, the velocity of the incoming primary extinguishes, and the re-emission is now propagating at c in relation to the refractive index of that medium.

Let's visualize a scenario. A source of light is approaching you. The incoming light from that source is traveling $c + v$ relative to you. Meaning, that particular light is traveling 299,792,458 m per second PLUS the speed of the source itself traveling toward you. The incoming primary light is absorbed, and a secondary light is re-emitted at c but with a refractive index of $n$. That's, $c$ divided by $n$, or $c/n$.

The so-called "Fresnel dragging"/ index of refraction of media is correctly expressed as a function of **frequency** $n\,(v)$, **NOT** wavelength $n\,(\lambda)$ as erroneously taught in textbooks.

So, from the *Extinction Shift Principle* under Galilean variance we get:

$$\frac{c+v}{\lambda_{c+v}} \text{ Before Interfering Medium} = \frac{\left(\frac{c}{n}\right)}{\lambda_n} \text{ Within Medium} = \frac{c}{\lambda_c} \text{ Re-emitted from Medium}$$

It doesn't matter if the medium encasing you is one millimeter thick or one mile thick. The re-emissions are in phase with the preceding emission.

The process of absorption and re-emission itself happens at the velocity of c, and the act of measurement introduces interference which produces re-emitted secondaries; therefore, it is physically impossible to ever measure the velocity of a primary emission from any given source. The velocity of light **IS** indeed dependent upon the velocity of the light source. But our own electrons fool us by absorbing the incoming primary which becomes extinguished/ extinct. Then those same electrons subsequently reproduce a new shifted packet of electromagnetic radiation from our own frame of reference.

Thus, Dr. Edward Dowdye's "*Extinction Shift Principle*" and a variable speed of light.

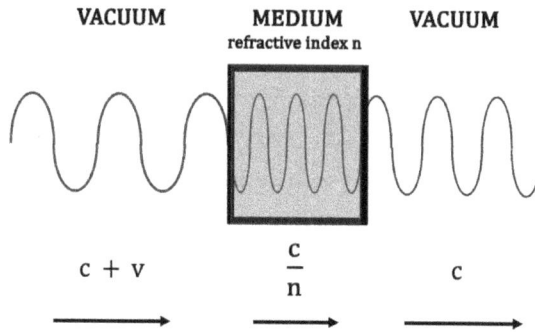

Fig. 2.10

In this image there is a sine wave propagating quickly from left to right. Under the sine wave says "c $+$ v" with thin arrow pointing to the right underneath the c $+$ v.

Above the sine wave is a caption in all capital letters saying VACUUM.

This indicates the velocity of light is dependent upon the velocity of the light source, and there is a light source moving out of frame of the image from left to right, quickly.

The approaching source from the left side of the canvas to the right justifies the motion of the sine wave to move faster since it is moving with velocity C plus the velocity of the source itself. The sine wave encounters a shaded rectangle. Indicating a refracting medium of some kind.

Below the shaded rectangle says $\frac{c}{n}$ with a thin arrow pointing from left to right situated underneath it. The sine wave has shifted to a higher frequency as the number of peaks and troughs increased the moment the sine wave hit the shaded box. The sine wave seems to slow down within the shaded box because the number of peaks and troughs increases, they are proportionally related. The shift in frequencies IS a change in the velocity of light.

The point at which the sine waves interact at the boundary of the shaded rectangle is in synchronization. The velocity of the wave has reduced and is re-emitted at the velocity of C, subject from to the refractive index of the shaded medium which acts as a window.

A brand-new sine wave exists the shaded box to the right and travels to the right. The number of peaks and troughs reduces slightly indicating the electrons making up the surface of the refracting medium re-emitted the new sine wave. The third sine wave travels to the right within the image. The portion of the image outside the boundary of the shaded box represents undisturbed vacuum of free space again. The liberated energy changes frequency once again and now propagates as a sine wave at the velocity of C relative to the shaded box.

So, the sine wave/ beam comes in from left and travels to the right within a vacuum, propagating at c + v ... hits a refracting medium... that primary is absorbed, and a secondary sine wave is re-emitted which shifts its frequency and velocity relative to within that medium. The moving sine wave then seemingly exists the medium to the right but is really re-emitted by the boundary of the shaded medium at the velocity of C within a vacuum again.

Since the frequency, $V$ doesn't change, the Energy, $hv$ must be conserved:

$$hv \text{ Before Interfering Medium} = hv \text{ Within Medium} = hv \text{ Re-emitted}$$

The measured value of the wavelength and the re-emitted/ Extinction Shifted wavelength of a packet emitted from an approaching source is expressed as:

$$\lambda_c = \lambda_{c+v}\left(1 + \frac{v}{c}\right)^{-1}$$

Photons are **NOT** particles. In _this_ emissions theory... there is no particle-wave duality. Light does NOT behave like bullets. But, if you were to shoot a bullet in microgravity, it would TRACE a straight-lined/ rectilinear path just as a collimated beam follows a straight-lined path. Also, like how the bullet from a gun will travel relative to the gun in came from... light also propagates relative to the bulb or source it comes from.

Wolfgang Pauli's ballistic principle applies in the regard that the initial wavefront of a given packet will propagate in a straight-lined path from a source to a target. Over long distances, that packet will have a delay at the velocity of light, C. The packet propagates relative to the source it comes from. If that source is traveling a given rate, you must add or subtract that rate on top of the velocity of light itself.... adding to the velocity of light C if the source is approaching towards the target and subtracting from the velocity of C if the source is receding away from the target.

We must always be mindful that the electrons making up the target itself will absorb that primary and re-emit a secondary which then travels relative to those electrons from that point on. So, even though the incoming packet might be approaching at $c + v$, the target will measure its own re-emitted light at C. There is a distinction between the velocity of light being _constant_ from its source versus the concept of the _constancy_ of the velocity of light being the same in _ALL_ frames of reference. The "_velocity of light constant_", C, is different than "_the constancy of the velocity of light_," especially in _all_ frames of reference.

Imagine a person with a gun. The person is standing on a train. The train is traveling by a grassy hill. There is a target on the hill. The bullet from the gun will hit the target at a higher speed if the person shoots that bullet while traveling on the train towards the target. The bullet will hit the target at $(\text{Bullet} + v)$.

You must account for the speed of the bullet towards the target PLUS the velocity of the train itself towards the target. If the bullet left the gun at the speed of light, the bullet would travel $c$ relative to the gun, but $c + v$ relative to target.

Target is 1 light minute away when the bullet leaves the gun.
Bullet travels the speed of light, c, relative to the gun.
The train is traveling 1/2 c toward the target.
How long will it take for the bullet to reach the target?
(30 Seconds)
Because the bullet is traveling c + v relative to the target.
The speed of light PLUS the speed of the train approaching the target.

The bullet travels c relative to the gun... but c + v relative to target.

1 Light Minute

The same applies to the boundary of a spherically expanding wavepacket/ photon/ graviton.
The edge of that expanding sphere grows at the velocity of light. If the gun were a LASER pointer... the wave-front of the beam will propagate at c + v relative to an approaching target. Or c - v relative to a receding target.

Fig. 2.11

The bullet travels out of the gun at the same speed regardless of if the gun is on a train or on the ground. But, if the gun itself is traveling towards the target... you must account for *that* speed as well *relative to the target.* The constancy of the bullet is NOT the same in all frames of reference. The bullet will indeed hit the target at a faster or slower speed depending on if the gun itself is traveling towards or away from the target.

That is an example of the ballistic principle and something that can be physically measured and observed. The "rest frame" of the bullet and gun is different than the frame of reference of the target it shoots at from the moving train.

Now let's imagine the target is 1 light minute away from the person holding the gun.

The bullet travels at light speed, $C$ relative to the gun. The train is traveling half the speed

of light, $\frac{1}{2} C$ towards the target. How long will it take for the bullet to reach the target?

(30 Seconds) The bullet will reach the target in 30 seconds because the bullet is traveling

faster relative to the target. In this case the velocity, $v = \frac{1}{2}$ the speed of light.

$$c + \tfrac{1}{2} c$$

The speed of light PLUS the seed of the train approaching the target.

The bullet travels $C$ relative to the gun, but $c + v$ relative to the target.

***All the light we ever see is the re-emitted light from the electrons making up ourselves.***
Since our own electrons encasing the atoms making up our physical bodies will absorb any incoming light. All light and electromagnetic radiation must pass through that "electron cloud" encasing our atoms. And by the time that incoming light ever reaches us... our own electrons extinguish it and generate a brand-new light.

I will be reiterating that point multiple times throughout this book.
This not only presents an alternative to general and special relativity, but also replaces Doppler Shift theory, aether theories and **ALL** past emissions theories like those from Wolfgang Pauli and Walther Ritz.

The *Extinction Shift Principle* is something new and unlike any previous model! This should reinvigorate the scientific community to take another look at **ALL** past experiments ALL past empirical evidence and data... and reinterpret how the end results arise for ALL past observations... Under "a new light" (pun intended.) The constancy of the velocity of light in ALL frames of reference is the cornerstone of Relativity. The entire postulate of relativity hinges on light being the same no matter what. If that's wrong... bye Relativity!

Fig. 2.12

Relativity was haphazardly built upon a foundation of sand. The introduction of Galilean variance is like a 9.0 earthquake rocking the scientific community. Liquefaction will win.

Fig. 2.13

# Wave-Fronts & Wave-Ends

Here are some more scenarios to think critically about. This model is in opposition to relativity's interpretation of how an observer would witness the same scenarios.

In this model, however, there's no need for mental gymnastics or billions of dollars of Hollywood CGI. You could work this out using software like Blender or just visualize it in your mind's eye... as long as you don't suffer from aphantasia.

So, to reiterate the main point:

Fig. 2.14

• We are **NOT** looking back in time we see galaxies, stars and distant locations.

• Light is **NOT** a recording of reality that replays the history of an event from that location.

• Reality is **NOT** broadcast.

• Reality does **NOT** dilate.

• Time is **NOT** relative as Einstein suggested even though signals indeed have a delay.

The entire theory of relativity is outdated with egregious misconceptions.
That model is based upon the framework of Lorentz invariance. We will be interpreting time and the perceptions of reality under a variable model for light instead. We will be confronted with the consequences of the Galilean variant interpretation. Every experiment, observation and test attributed to relativity can be reinterpreted to dismantle relatively. Every... single... one. So, let's begin with some hypothetical scenarios of how we can view light, time and reality. But first we must talk about how light propagates. Then we'll apply that to gravitation later on.

# What is a "Wave-Front" & What is a "Wave-End"?

I'll start by showing water coming out of a hose as an analogy for light coming out of a laser pointer. The arrow indicating where the water first starts coming out of the hose is the front part of the wave... the Wave-Front.

Fig. 2.15

Where the water stops coming out of the nozzle is the Wave-End... where the water ceases.

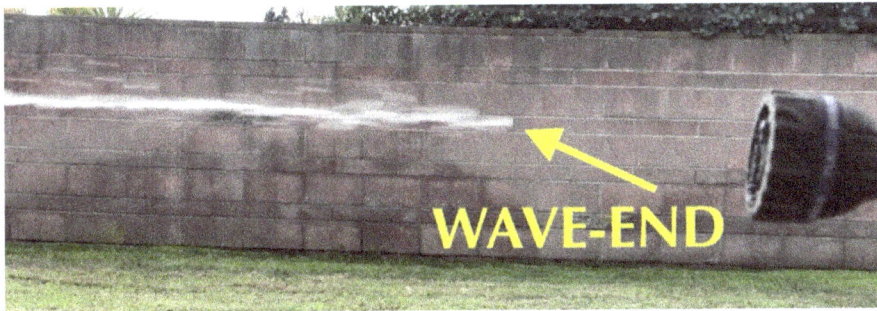

Fig. 2.16

Take note, the water that already left the hose will continue to propagate out even if the hose is turned off or is destroyed.

And if the hose or the nozzle itself is traveling, you must account for *that* speed in addition to the speed at which the water is coming out of the nozzle.

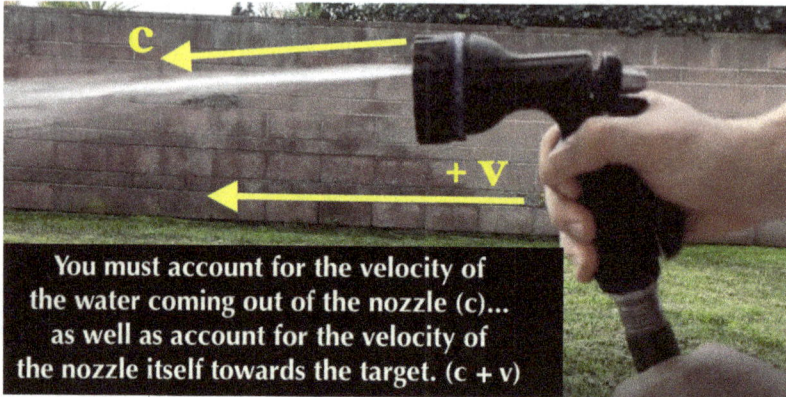

Fig. 2.17

Now imagine the hose is a laser pointer, and the water shooting out is light traveling at light speed. If the hose or the laser pointer itself is moving in a particular direction, then you need to account for that and add *that* **on top** of the speed of light.

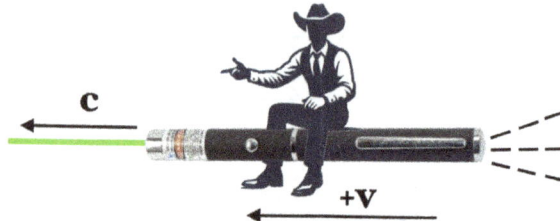

Fig. 2.18

The water comes out of the hose at a constant rate. The light comes out of a laser pointer at the rate of C, but if the source itself is moving, then you must *add that* velocity *on top* of the light. So that means, the velocity of light *is indeed* dependent upon the velocity of the light source. But there's an illusion. And the variant Galilean alternative to the Lorentzian invariance of the wave equation, derived by Dr. Edward Dowdye as shown at the beginning of the book, mathematically exposes that illusion.

## Convex Wave-Front

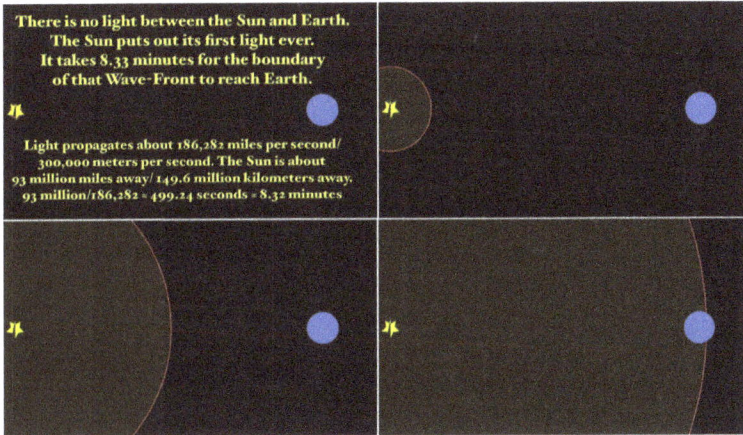

There is no light between the Sun and Earth. The Sun puts out its first light ever. It takes 8.33 minutes for the boundary of that Wave-Front to reach Earth.

Light propagates about 186,282 miles per second/ 300,000 meters per second. The Sun is about 93 million miles away/ 149.6 million kilometers away. 93 million/186,282 = 499.24 seconds = 8.32 minutes

Fig. 2.19

## Concave Wave-End

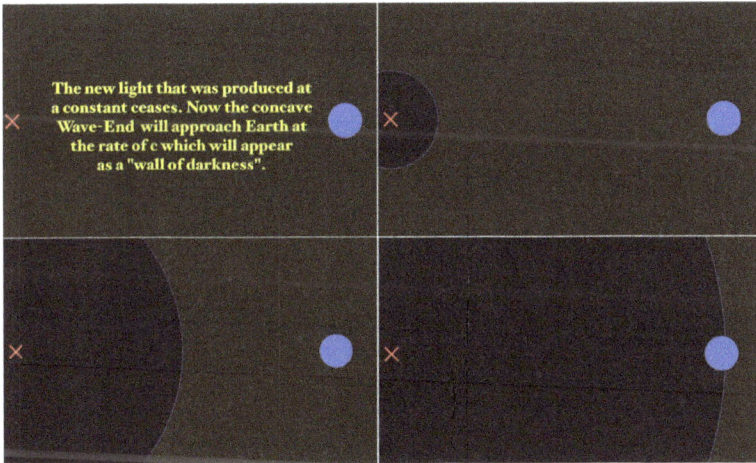

The new light that was produced at a constant ceases. Now the concave Wave-End will approach Earth at the rate of c which will appear as a "wall of darkness".

Fig. 2.20

Both convex Wave-Front and concave Wave-End propagate at the rate of C. This type of wavefront is called a "spherical wavefront." (by the time the boundary of the wavefront reaches Earth though, the size of the wavefront is so large, it acts like a plane wavefront.)

Einstein thought the velocity of light is the same to everyone in *all* frames of reference because whenever an observer measures, that's all that any observer can ever detect. Since the light re-emitted by the electrons making up yourself is traveling relative to yourself, it serves the illusion that the velocity of *all* light is the same in *all* frames. Because the observer, again, assumes he is witnessing the same primary light from the source he's looking at. That primary light could be coming in at a faster velocity or slower velocity depending upon that source approaching towards you or receding away from you. You can never discern the velocity or the wavelength of the primary from *any* source because the electrons making up the observer absorbs that incoming primary and re-emits a brand-new secondary.

There are illusions within illusions. Like a labyrinth or hall of mirrors within a hall of mirrors. Get lost in a reflection and you'll run face-first into a wall. The scientific community constantly runs full speed into walls like a fly futilely trying to pass through a glass window. And then repeats actions expecting different results. Where have we heard that one before? How ironic.

Albert Fly-nstein

Fig. 2.21

# Magical Gun/ Bullet Re-Emission Analogy

Imagine a magical gun that can absorb any incoming bullet that hits it. This magical gun automatically fires its own bullet every time it absorbs an incoming bullet. It doesn't matter the caliber or velocity of the incoming bullet. Whenever it absorbs an incoming bullet, it will eject and shoot out its own bullet at its own caliber and velocity.

The owner of the gun thinks ALL bullets have the same caliber and velocity as his own gun. This is because the owner of this magical gun is only able to measure the bullets from his own gun. In reality... the barrel length is different on various guns... the caliber and velocity of incoming bullets will vary drastically. Yet ANY time the owner of the magical gun measures the caliber and velocity... he is only able to measure the RE-EMITTED bullet from his own gun. Serving the illusion that ALL bullets have the same caliber and velocity.

Even if someone is shooting a gun while on a train heading toward him... doesn't matter the incoming velocity of the bullet.... the gun absorbs it and shoots out its own bullet at the velocity of C. Therefore, the owner of the magical gun thinks ALL bullets travel at the velocity of C. Any time this guy picks up a new gun... it absorbs all incoming bullets and converts them to its own caliber and velocity.

How would this theoretical gun own ever be able to tell the true velocity of an incoming bullet? He couldn't. It would be impossible for this hypothetical person to ever measure anything else. Because the act of measurement generates a new bullet. All the bullets this observer can measure are the re-emitted bullets from his own gun; just as, it's impossible for ANY observer to measure the velocity of light as anything other than C. This is because the electrons making up the observer will absorb all incoming primary light and all packets of electromagnetic radiation... those primaries will diminish to zero, while those same electrons re-emit a brand-new secondary packet as an equal and opposite reaction. That process happens at the velocity of C, serving the illusion the primary packet distorted or "reflected." It is a brand-new light/ brand-new electromagnetic radiation.

This concept follows with Wolfgang Pauli's ballistic principle of light, but the flaw with that model is the assumption that light itself is made of particles, which light is not. That is a misinterpretation rooted in  the double slit experiment which I'll also address. It's an illusion.

All observers are only capable of measuring the re-emitted light from the electrons making up themselves. But photons are not bullets or particles at all. There is no particle wave-duality or a literal mass to energy conversion. The double slit experiment is misconceived. The same laws of Fermat apply to the tangential re-emissions of packets as the angle of incidence is equal to the angle of reflection.... but so-called "reflections" are really secondary re-emissions of light being projected at a tangent equal to the angle of incidence/ absorption of the incoming packet. And the laws of Hamilton and Newton apply to the particle aspects of the experiments.

The same ballistic laws of physics apply to light in the regard that the wavefront of a given packet will be propagating at the velocity of c PLUS the speed of the source traveling toward or away from the observer. The angle of reflection/deflection of a beam of light is the same path that a particle would follow in a straight line/ rectilinear path in undisturbed space. Then describes the ricocheting in a straight line equal to the angle of incidence/ approach of that particle.

This means the light from a particular source will be traveling $c +$ the speed of the source $(v)$ or $c -$ the speed of the source $(v)$. But again, the illusion is that by the time ANY primary light reaches the observer, that primary is absorbed by the electrons making up that observer resulting in a secondary light altogether being re-emitted buy those same electrons. The electrons of the observer travel relative to that observer, serving the illusion that *all* packets of electromagnetic radiation propagate at the velocity of C in *all* frames.

The electrons making up yourself will absorb any incoming light and then shoot out its own light. You can only measure the re-emitted light shot out by your own electrons... AFTER they absorb energy from incoming light.

# Magical Golf Ball Re-Emission Analogy

Any time the golf ball hits something, it disappears and is absorbed into whatever it hit. But simultaneously, a new golf ball emerges right in the same spot and then "ricochets" off. (Photons are Not ball particles... this is just an analogy for absorption of primaries and re-emissions of secondaries.) When you putt the golf ball it goes in a straight line. But as soon as it hits a wall... the wall absorbs it... and a brand-new golf ball is emitted from that same part of the wall. (Making it seem like the same ball is bouncing back.)

Every time the ball breaks it's straight path, a brand-new ball takes its place. The ball is replaced at the velocity of light, **C.**

(The ball in this case represents the center of an expanding sphere/ an emission of light.)

Mainstream science and the big bang model suggest the same light has existed since the beginning of time and that same light just bounces around over and over until it reaches us. But when you see a reflection of yourself in the mirror... that is not the same light. The same light does not bounce and reflect off of the glass. The electrons making up the surface of the glass absorbs the incoming light and then a brand-new secondary light is re-emitted from the glass. That happens so fast it *seems* like the same light bounces off.

Now imagine a game of pong where a ball is bouncing off of a moving wall.
The ball is absorbed by the moving wall and a new ball will emerge from that same point. That new ball will be traveling the same speed as the moving wall when it is re-emitted and traveling from the frame of reference of the wall itself. Imagine an expanding sphere emanating and growing from the center point of that ball. The ball is traveling relative to the wall it was re-emitted from in addition to a spherical burst propagating outward at the rate of C from the center point of that re-emitted ball. So, the center point of that spherical wavefront is traveling away from the wall yet traveling relative with the wall itself. Like if you jumped off a train. You are moving relative with the train. Jump on to another train... you are now traveling relative with that train. Each time you jump off you are at a new speed. Not that you stay the same speed and space-time warps instead. That's absurd.

— CHAPTER 3 —

# Walter Bowman Russell

Back in the 1920's, a man named Walter Russell tried to communicate the idea that light re-emits when he said, "Light doesn't travel." That statement and sentiment was repeated in his books, *The Secret of Light* and *The Universal One,* " which he rewrote and released as *A New Concept of the Universe* in 1926. The statement that light doesn't travel was taken the wrong way. Russell stated, "neither light nor heat flows from one point of space to another." He stated, "the light and heat which appear to come from the star or the sun has never left the star or the sun. That which man sees as light and feels as heat is the reproduced counterpart of the light and of the heat which is its cause."

Walter Russell said that wavepackets do not travel, but rather are "reproduced from wave-field to wave-field." Which I think is a more esoteric way of saying the SAME packets do not reach you, but rather are re-emitted from medium to medium. The light is reproduced as secondaries at points of interference. It's never the same light, never the same gravity, or electromagnetic radiation. There is no *constancy* in *all* frames of reference.

**Walter Russell**

Fig. 3.1

# Lifting the Veil of Illusions

Everything from the color of prisms to the double slit experiment has been interpreted by people fooled by illusions. I have my own heretical takes, and I'm not fond of a lot of quantum theory. I think of it as "Quantum Quackery," as the claims are founded upon an invariant model of light. It's resulted in a century's worth of misconceptions

Critical thought has been deferred to "the experts" who have dismissed simpler explanations for the same scenarios. The result has produced conclusions built upon ad hoc

Fig. 3.2

assumptions and illusions of observation interpreted under Lorentz transformations.

Lorentz invariant transformations are like rose colored glasses, giving a false filter of reality. Like the movie *Dr. Strange in the Multiverse of Madness,* when Strange firsts visits the Scarlet Witch on a beautiful orchard. Dr. Strange soon recognizes the artifice and illusion for what it really is. When the veil is lifted, he's able to see the decimated world around him.

Galilean transformations are like the fictional glasses from the movie "They Live". They allow you to lift the veil of illusions of Lorentz transformations. With the proper fundamentals, people can see what the last 100 years of so called "modern" science of incoherence has done to the world... and it ain't pretty. The scientific community thinks relativity has done a great job, and they pat themselves on the back touting their "achievements," not realizing that route is leading to our own destruction and demise... on *our* own dime! All while they fantasize about science fiction scripts which ironically are ultimately achievable... but only by dropping the very relativity they think will achieve it. I guess Einstein had an Oedipus complex... but it was with his cousin instead.

Fig. 3.3

As for the double slit experiment, there are four different scenarios. But many people are accustomed to think there is only one scenario that magically changes between two options based upon observation. As if objective reality itself is manifested by a subjective observer. By that arrogant logic, the universe shouldn't exist to blind people.

Their sanctimonious high horse is about to get a broken leg.
Is that.... wait... is that, Tonya Harding? Oh no...

Fig. 3.4

It's like the old adage, "If a tree falls in the woods and no one is around to hear it... does it really make a sound?" Of course it does. Since the definition of sound is the vibration or displacement of air at a given rate/frequency. Trees require air to live. If a tree falls, it must displace that air and move it out of the way as it falls. Just because a human isn't around to interpret the particular difference of compression and rarefaction of the air doesn't mean it isn't taking place in totality. The universe doesn't emerge or manifest or revolve around a given subjective observer.

Boil one gallon of water at sea level in an aluminum pot. It doesn't matter your feelings or subjective observation, the water will boil at the same rate, given the variables are the same, regardless or not of who watches it. They say a watched pot never boils... well that's not true. I've seen it for myself! Just requires patience within objective reality.

Not everything we are taught or told is accurate. When looking at things for yourself, like a prism and colors for example, one can see something very different than what textbooks claim. Psychologically though, people tend to violently defend the first model they are brought up with. The first scientific model engrained into the mind of a child will be branded in that child's mind. So, if you introduce relativity to people at a young age... it will be that much more difficult for them to drop it later on. Relativity is what the most trusted people in their lives have presented to them from the start. "So you're telling me everyone is wrong then?!?" YES... I am.

As Alvin Toffler said, "The illiterate of the 21st century will not be those who cannot read and write, but those who cannot learn, unlearn, and relearn." And as Sir Ken Robinson suggested in a TED talk called "Do Schools Kill Creativity," we unfortunately have a school system now adays which has replaced an education system. "Schooling" just cranks out regurgitators and people who can memorize and parrot. The masses of human bots and NPCs. Just smart enough to push a button but not smart enough to ask what the button does, what's it made of, or who built it and why. "Educating" teaches people *how* to think while supplying them with all the information on the table to sort through... not just telling people *what* to think and testing one's ability to regurgitate it.

It is crucial for the scientific community to distinguish between the limitations of our senses and the influences on a given sensor or measuring apparatus. There is a discernment to be made between the interference *we* produce by the very measuring devices we use versus the intrinsic nature of the phenomenon we aim to measure. And that discernment has nothing to do with some non-existent "particle-wave duality" invented by QED.

For years, I thought the double slit experiment was about sleeping with two women. But apparently, nerds all around the world take it to mean something about shining light through gaps in a wall. Imagine my surprise to learn about optics and quantum theory. At least my original experiments weren't totally in vain.

# The Double Slit Experiment(s)

Here are 4 individual, definitive and objective scenarios for the double slit experiment(s). Plural!

(i)   You shine a spherical wavefront of light through the slits.

(ii)  You shine a plane wavefront of light through the slits.

(iii) You shoot electrons wildly from a hot source through the slits.

(iv)  You shoot electrons from a hot source & focus them in a beam through the slits.

When you block a plane (flat) wavefront, it produces spherical (convex) wavefronts just like Huygens' principle. The sharp edges of the slits re-emit cylindrical wavefronts. The sharp edges of the slits also emit secondary electrons from an initial impact of a primary electron from the focused and collimated beam.

In order to produce the interference patterns, you have to manipulate the setup to have the double slits fit inside of the diameter of the beam you are shooting.
If you use a red laser pointer, the diameter of that beam (the beam waist) is about 5mm. So you must have both slits fit within the width of that 5mm beam. Then, of course you're going to get an interference pattern. But if you block a slit, you won't get same interference pattern.

You also have to put both slits within the diameter of the electron beam which is about 5μm and the slits themselves must be about 100 nanometers in order to fit in that beam. The primary electrons from the beam impact the 4 sharp edges of the slits and re-emit secondary electrons which form the pattern on the screen over time. Of course you're going to get a pattern. But they physically block one of the slits with a sensor to prevent the formation of the pattern and they call that "an observer". Of course you're not going to get the same pattern because you're blocking the stream of particles from getting to the screen. Change the parameters of the setup, you change the results.

There is no mystery or paradox or particle-wave duality needed for the interpretations. All the same mysteries can be explained using classical physics, Hamilton's equations, Galilean variance, optics, electrical engineering and logic.

Electrons absorb, emit, and re-emit all light and all electromagnetic radiation. All atoms contain electrons. The wall with the two sharp-edged slits is made of atoms.

• Shine light through the slits.
You can shine a spherical wavefront through the slits and/or a plane wavefront through the slits. Not all the light seen on the wall is the same primary light shining from the source. Some of the primary light makes it through the slits undisturbed... But the electrons making up the wall with slits itself will absorb some of the incoming primary. Those same electrons re-emit a secondary packet as an equal and opposite reaction which is re-emitted from the 4 edges of the slits themselves. The wavepackets re-emit from the edges of the slits thereby producing cylindrical wavefronts. Some of those wavefronts overlap the primary light and contribute to the interference pattern.

• Shoot charged particles through the slits.
A 4000° filament shoots off electrons. The electrons are focused with a magnetic field to form a beam. The beam of particles is sent through the slits. The electrons from the beam hit the edges of the slits. The primary electrons induce secondary electrons to eject from the atoms making up the edges of the slits. The particles deflect at specific angles and produce a pattern from the coherence of the beam. Without the magnetic field manipulating the particles in a beam, the particles produce 2 strips on the wall.

• Shine light and shoot charged particles through the slits.
You get the combination of dots forming in a wave-like distribution in the last explanation but also the light from that same tungsten bulb produces 2 bands on the screen. But no one is concerned with the pattern of light in that experiment since they are only after the pattern produced by particles over time. But the light from the 4000° tungsten filament/ bulb produces 2 bands because that light is a spherical wavefront. So you get a distribution of particles overtime, but 2 stripes of light during the experiment. Has nothing to do with a literal flipping of mass to energy or that particles themselves *are* the waves as well.

Now place a detector in the path, and the electrons making up the detector will absorb that primary and re-emit a brand-new secondary packet which is measured as the frequency of interference between the source and itself. That new packet is now traveling relative to the electrons making itself.

You can manipulate the setup to produce 4 distinct combinations of results from 4 different experiments. NOT that observation magically switches the states.

(i)     2 bands of light and 2 bands of particles.

(ii)    Interference pattern of light and an interference pattern of particles.

(iii)   Interference pattern of light and 2 bands of particles.

(iv)    2 bands of light and an interference pattern of particles.

(i) The 2 bands of light and 2 bands of particles results from the $4000°$ thermionic emissions from the tungsten filament sent through the slits *without* focusing them into a beam. The particles shoot wildly, analogous to a cannon full of marbles. Unrelated to the emissions of particles, the spherical wavefronts of light from the same filament simultaneously propagate through the 2 slits. The Spherical wavefronts produce 2 bands. (you would need some sort of screen detector that can simultaneously pick up the impression of light as well as the accumulation of particles over time.)

(ii) The interference pattern of light and interference pattern of particles would result from two different emitters. A thermionic emission source shooting out electrons that are focused into a collimated beam using a magnetic lens. The filament shoots out electrons but not visible light. Any light from that filament would produce spherical wavefronts. We want to avoid that. We would need to use something like a filament of lanthanum hexaboride (LaB6) which mainly produces infrared light under the right current and load. So it would produce electron emissions but not visible light. A photosensitive screen/ detector screen specific to visible light could be used. In addition to the LaB6 filament and magnetic lens producing an electron beam... a second light source like a laser would be aligned to point through the same slits as the electron beam. The photosensitive screen would pick up the interference fringes from the coherent light *and* simultaneously pick up the accumulation of particles in an interference pattern over time.

(iii)  An interference pattern of light and 2 bands of particles would result from two different emitters. A thermionic emission source shooting out electrons wildly that are *not* focused into a collimated beam. Just the LaB6 filament producing electron emissions analogous to marbles in a cannon and that same filament producing infrared light. But again, the photosensitive screen only picks up visible light and the accumulation of electrons over time. Unrelated to the electron emission source, there is a second light source like a laser which is aligned to point through the same slits as the electrons from the LaB6 filament. The photosensitive screen would pick up the interference fringes from the coherent light and simultaneously pick up the accumulation of particles in 2 bands.

(iv) The 2 bands of light and an interference pattern of particles would result from the 4000° tungsten filament producing electron emissions. Those electron emissions are focused into a collimated beam using a magnetic lens and then sent through the 2 slits. The light from the same filament shines through the same 2 slits. Because the light is a spherical wavefront, it produces 2 bands on the photosensitive screen.

There are more scenarios.

(v)     2 bands of light AND an interference pattern of light.
(vi)    2 bands of particles AND an interference pattern of particles.

(v) 2 bands of light and an interference pattern of light results from overlapping a coherent light source (plane wavefront) and an incoherent light source (spherical wavefront). Two different light sources are simultaneously sent through the same slits.

(vi) 2 bands of particles and an interference pattern of particles results from overlapping the emissions produced from a collimated electron beam from a thermionic Source A with a second thermionic Source B, without a focused beam. The emissions are simultaneously sent through the same slits and overlap to produce both two bands on the wall and interference fringes.

And more scenarios.

(vii)   2 bands of light, 2 bands of particles AND an interference pattern of light.

(viii)  2 bands of light, 2 bands of particles AND an interference pattern of particles.

(ix)    2 bands of particles, an interference pattern of particles AND an interference pattern of light.

(x)     2 bands of light, an interference pattern of light AND an interference pattern of particles.

(vii) 2 bands of light, 2 bands of particles AND an interference pattern of light results from a thermionic heat source and an additional coherent light source. The light from the glowing 4000° tungsten filament is sent through the slits to produce 2 bands of light while the particles wildly emitted from that same filament produces 2 bands of particles. The light from the filament produces a spherical wavefront therefore you get 2 bands of light in the same place as the 2 bands/ strips of particles. But in addition to all that, there is a second light source like a laser. And the light from the filament overlaps the light from the laser beam through the same 2 slits which are situated within the diameter of the laser.

(viii) 2 bands of light, 2 bands of particles AND an interference pattern of particles results from two separate thermionic heat sources, A and B. One of them has its emissions focused into a collimated beam and the other does not. Thermionic heat Source B has its electrons emitting off wildly. The slits are within the collimated beam from the thermionic Source A. That will produce a simultaneous 2 bands of particles AND an interference pattern of particles. The light from both thermionic heat sources A and B will produce 2 bands of light since the light is spherical wavefronts in this scenario. If you were able to turn on/off source A or B, you would see the patterns change only because you turned off one of the thermionic heat sources. Not because they magically change back and forth from particles to waves.

(ix) 2 bands of particles, an interference pattern of particles AND an interference pattern of light results from two different thermionic heat sources, A and B. One of them has its emissions focused into a collimated beam and the other does not. Thermionic heat source B has its electrons emitting off wildly. The slits are within the collimated beam from the thermionic source A. That will produce a simultaneous 2 bands of particles AND an interference pattern of particles. But the heat sources are using LaB6 filaments   specific to mainly infrared as to avoid making a pattern on the screen within the visible light spectrum. There is an additional coherent light source like a laser which points at the same slits where the electrons from the two thermionic heat sources are shot through. This will produce the two different particle patterns AND an interference pattern of light in the visible spectrum. If you toggle On/Off between thermionic heat source A, B and the laser, you can switch between patterns observed. Again, demonstrating no literal particle-wave duality or flipping between mass and energy.

(x) 2 bands of light, an interference pattern of light AND an interference pattern of particles results from two different light sources, one of them being a thermionic heat source and the other a laser. The emissions from the thermionic heat source using a tungsten filament are focused to a collimated beam. In addition to that, there is another light source like a laser pointing through the same slits as the collimated electron beam. That will produce 2 bands of light from the tungsten bulb overlapping the interference pattern of light from the laser AND an interference pattern of particles accumulating over time on the detector screen.

There are 15 total combinations. The others are just isolating each single variable. There are really 16, but the 16th is just the lack of all the variables. So nothing at all.

(xi)     2 bands of light only.
(xii)    2 bands of particles only.
(xiii)   An interference pattern of light only.
(xiv)    An interference pattern of particles only.

(xi) 2 bands of light only is accomplished by sending a spherical wavefront of light through any sized slits. Cut 2 slits in some paper. Shine your flashlight from your phone through the slits. Produces 2 bands of light. Even the glow from a coherent laser beam will produce 2 bands of light. It's only if the 2 slits are small enough to fit within the diameter of the laser beam itself do you get an interference pattern from the laser.

(A great resource for learning about lasers is YouTube user styropyro)

https://www.youtube.com/@styropyro

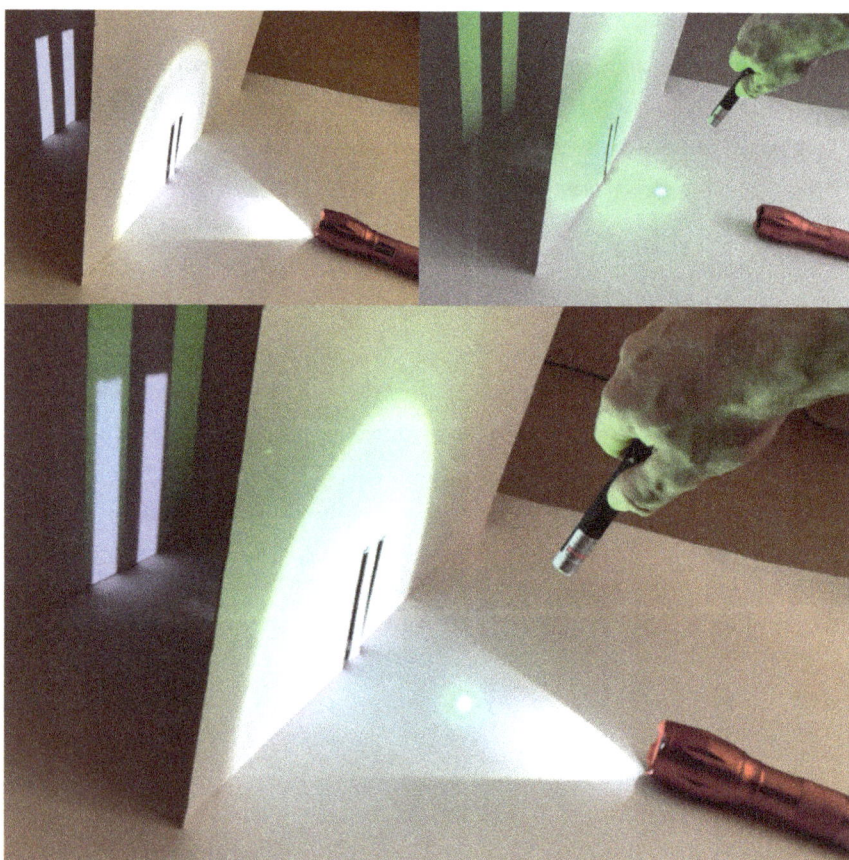

Fig. 3.5

(xii) 2 bands of particles only would only be accomplished with the LaB6 filament which wouldn't emit visible light, yet the filament would produce thermionic emissions as wildly shooting particles.

(xiii) An interference pattern of light results from the 2 slits being within the diameter of a collimated/coherent beam. The beam is a plane wavefront. You can put a small 18 gauge wire over the end of a laser pointer, right in the middle of where the beam comes out. The wire will split the beam. Then you tape the wire over the end of the laser pointer where the straight edge of the piece of tape is aligned parallel right next to the wire on either side. It must be precise, but anyone with a laser pointer can try. Fig. 3.6 is a crude example you can see in the text version of the book. It's difficult to take a picture of the interference pattern with my iPhone 8 camera, but in person, you can see the strips of light and the fringes.

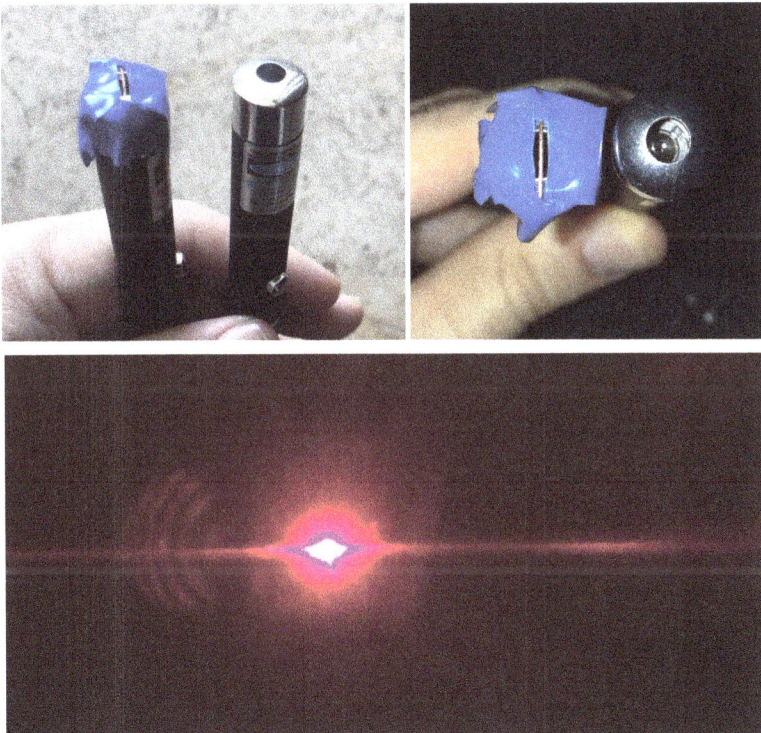

Fig. 3.6

(xiv) An interference pattern of particles would only be accomplished with the LaB6 filament which doesn't emit visible light, yet the filament would produce thermionic emissions shooting off particles. The emissions would need to be focused using a magnetic lens and then the collimated beam pointed at the 2 slits. The 2 slits need to be within the diameter of the beam.

Then, all of the scenarios simultaneously. Imagine a setup where all the factors are present and you can turn on and off each variable to produce any of the given 16 possible combinations.

(xv)    2 bands of light, 2 bands of particles, an interference pattern of light AND an interference pattern of particles.

(xvi)   No bands of light or particles, No interference patterns of light or particles at all.

(xv) 2 bands of light, 2 bands of particles, an interference of light AND an interference of particles results from having two different thermionic heat sources A and B. Thermionic heat source (THS) A has a tungsten filament with its emissions focused into a collimated beam and THS B does not. The light from the filament produces spherical wavefronts. In addition to all that, there is a coherent light source where the laser beam is pointed through the same slits as the collimated beam from THS A. The light from the laser beam produces plane wavefronts. So you get the light from the filament and the light from the laser overlapping to produce two different patterns... 2 bands and an interference pattern. This arrangement also simultaneously produces 2 bands from particle emissions and an interference pattern from particle emissions.

(xvii)  The absence of all factors. Nothing is turned on which produces no result at all.

There are 16 possible combinations in 5 groups:

1. No variables present:
   - none
2. One variable present:
   - A
   - B
   - C
   - D
3. Two variables present:
   - A, B
   - A, C
   - A, D
   - B, C
   - B, D
   - C, D
4. Three variables present:
   - A, B, C
   - A, B, D
   - A, C, D
   - B, C, D
5. All four variables present:
   - A, B, C, D

None of these combinations or scenarios involve a literal mass to energy interchange nor a particle-wave duality. The clicks of the double slit sensor are the particles being prevented from making it to the screen. Which means there isn't the same pattern. The double slit experiments have nothing to do with a real-world human observing the scenario as if the patterns switch as he blinks or looks away. That's an egregious misnomer. People think there's just one scenario flipping back and forth because "experts" implied so in erroneous documentaries like *What the Bleep Do We Know, Flatland* or *Down the Rabbit Hole*.

People believe the misnomers because of science fiction nonsense peddled by sanctimonious know-it-alls like Neil deGrasse Tyson or Dr. Michio Kaku. And when you question people on any discrepancies, they cop out and chalk it off as quantum magic, as an oversimplification, or they claim, "the quantum realm is just too mysterious to comprehend." And if you think you know it, you don't... because it's not knowable... yet it is knowable at the same time because of superposition... but only the "experts" can be in that super position to know it... even though they admit they don't know what the fuck they're talking about. Then we have the people who slap the word "quantum" on the label of their products to garner more sales from suckers. Literal snake oil has more use than most all products claiming to have quantum effects (for only $19.95).

Funny enough, even "King Jellybean" Erwin Schrödinger thought the Copenhagen interpretation was nonsense. Schrödiddler made up the whole "cat in the box" thought experiment as a joke to make fun of the absurdity of quantum theory. But the world still took it as valid or literal and now use that as justification for the laughable belief that "all things are possible until you look." By that logic, burying your head in the sand will make all your problems go away. Schrödinger also hated probability theory and was quoted in a June 13th, 1946, letter to Einstein saying, "God knows I am no friend of probability theory, I have hated it from the first moment when our dear friend Max Born gave it birth. For it could be seen how easy and simple it made everything, in principle, everything ironed and the true problems concealed. Everybody must jump on the bandwagon."

| Aspect | Quantum Explanation | ESP Explanation |
|---|---|---|
| Nature of Light | Photons (particle-like) | Continuous electromagnetic waves |
| Energy Transfer | Photons transfer energy in discrete quanta | Waves induce oscillations and transfer energy continuously, potentially re-emitting at different frequencies |
| Threshold Condition | Photon energy must exceed work function | Accumulated wave energy must exceed work function |
| Kinetic Energy of Emitted Electrons | $K = h\nu - \phi$ | $K = h\nu - \phi$ (same equation, different interpretation) |
| Explanation Basis | Quantum mechanics | Classical wave theory |

Fig. 3.7

# Quantum Eraser Experiment and Delayed Choice

Classical physics dictates that if a scientific approach can yield and correctly predict all processes in physics, then that approach must be totally independent of the observer. In order to understand the claim, you have to understand what the experiment is set up to conduct and what actually takes place and what is actually measured. Need to isolate the assumptions and claims from what the function of the setup entails. The quantum eraser and delayed choice is claimed to happen ONLY with the thermionic heat source focused into a collimated beam. In that setup, ONLY the particles are being considered and that setup is specific for detecting clicks of particles.

Electron source *** Magnetic lens ===> Double slit ||| ===> Wall with Detector SCREEN where the electrons impact over time to show an interference pattern. But after the slits and before the detector screen on the wall, there is a sensor called a phototube. The impact of the electron generates an electromagnetic signal which then generates a click on a meter when the signal is generated. Like a Geiger counter. Let's call it a mini Geiger counter (mgc).

The clicks result from electrons hitting the mgc when it is placed by the slit in between the slit and the wall with the detector SCREEN. The mgc is a detector as well. So that might be confusing. There is a wall with the slits and a wall with the detector screen. There is a detector screen and a mini detector or the scintillator detector. Photons play NO role in this set up at all. So any talk of photons in the quantum eraser claim is bogus assumptions.

What is ACTUALLY taking place in the experiment though? A PARTICLE beam of electrons. And when the secondary electron emission are flying off the edges of those slits, the mgc generates a click as a response. Each click in the mgc is an electron that didn't make it to the detector SCREEN on the wall. Which means the mgc is BLOCKING the slit and preventing the path of secondary electron emissions from making it the detector SCREEN.

The claims made for the delayed choice and eraser are erroneous. They use the amount of clicks in the mgc to RECONSTRUCT the distribution pattern later. Like an overlay or side by side comparison. And that comparison shows a difference of 2 bands of particles or an interference of particles from the RECONSTRUCTION of the data. There is no real world observer. They use a damn Geiger counter clicker to map out the patterns that didn't make it to the screen. Thus the difference of 2 bands and interference pattern.

Then they interject wild claims about photons and electrons flipping back and forth and depict the scenario as if it's human sized and there's someone looking at the experiment between the slits and detector SCREEN. But it simply does not work like that at all. It's spurious and deceitful depictions. So the explanation under classical physics is a total rejection of the claims. Statistical mechanics and optics still apply. But the scenarios must be accurately isolated in the language describing results. Photons are not considered whatsoever in the quantum eraser and delayed choice experiment. That's an assumption. The photons are from a plane wavefront or a laser or PHOTON source. The quantum eraser and delayed choice claims are made for the THERMIONIC HEAT source.

If you're dealing with strictly photons (wavepackets), then you're dealing with classical wave mechanics, probabilities and optics. Electrons are not considered at all in that experiment. The only way electrons are considered is if a THERMIONIC HEAT SOURCE is ejecting electrons. That requires a magnetic lens to focus the electrons into a collimated beam LIKE a laser. But again the laser is a PHOTON SOURCE and the electron beam is a THERMIONIC HEAT SOURCE coming from a tungsten filament and high voltage transformer with a magnetic lens.

Photons are not considered in that set up at all. Electrons are not considered in the laser experiments at all. Two completely different experiments and setups to yield two completely different results. But people think they are one in the same experiment. There are FOUR different experiments and setups. The quantum eraser is claimed to combine two of them which have never been done. It's *either,* a THERMIONIC HEAT source (tungsten filament).... OR a PHOTON source (laser pointer).

Electrons are not considered in the laser pointer experiments and photons are not considered in the filament experiments. There are volumes of papers on the net to distinguish between the scenarios. You have to read what the apparatus is capable of in their experiment and then look at the claims made for the apparatus. How can they claim electrons present if they are only using laser beams? That's silly. That would be like using a Crookes tube and insisting there is no electrons present, it's only a laser beam. Need to distinguish what is ACTUALLY taking place and then separate that for the claims being made for that set up. Here are some derivations using Galilean variance and modified Hamiltonian equations as a purely classical interpretation of the double slit experiment:

The sharp edges of the slits produce cylindrical wavefronts since the wavefronts relatively adopt the shape of the emitter. The wave equation for the cylindrical wavefronts is:

$$\frac{\partial^2 \psi}{\partial r^2} + \frac{1}{r}\frac{\partial \psi}{\partial r} + \frac{\partial^2 \psi}{\partial z} - \frac{1}{v^2}\frac{\partial^2 \psi}{\partial t^2} = 0$$

The boundary condition for the re-emission of packets is:

$$\psi(r,t) = A\cos(kr - \omega t + \varphi)$$

The "superposition" of the total wavefield is:

$$\Psi = \sum_{i=1}^{4} A_i \cos(kr_i - \omega t + \varphi_i)$$

The "Hamiltonian" of the system/ total energy and sum of kinetic energy from motion is:

$$H = \int \left( \frac{1}{2}\varepsilon_0' \left(\frac{\partial \Psi}{\partial t}\right)^2 + \frac{1}{2}\frac{1}{\mu_0'}(\nabla\Psi)^2 \right) dV$$

The intensity of the interference is shown as:

$$I = |\Psi|^2 = \left| \sum_{i=1}^{4} A_i \cos(kr_i - \omega t + \varphi_i) \right|^2$$

The modified factors under Galilean variance are:

$$k' = k\left(1 + \frac{v}{c}\right) \quad \text{and} \quad \omega' = \left(1 + \frac{v}{c}\right)$$

# Michio Kaku and String Theory

I met Michio Kaku in 2008 at the Exploratorium in San Francisco. That was before the Exploratorium moved to its current location on Embarcadero St. from the Palace of Fine Arts area. I don't want to digress too much about the Exploratorium, but that was one of my absolute favorite places as a kid. A giant warehouse of hands-on scientific experiments. Frank Oppenheimer founded the Exploratorium. It's honestly one of the greatest places a kid or curious adult can possibly visit... let alone grow up around and have access to all the time. There's over 1000 different exhibits and cool things to try out and see. One of the coolest places. Your kid will LOVE it! Check it out and go visit if you're able.

Anyway... Michio Kaku was speaking there for his book tour for *Physics of the Impossible*. My dad took me. I was the first person in line for the meet and greet after the presentation. I was the only person to get a photo with him that day that I know of. (I ruined it for everyone else. haha). I asked to take the picture first before spending a good 7 or 8 minutes grilling Michio about different "free energy" technologies and some challenging alternative scientific models I had come across. This guy dismissed all of it and personally told me, "I looked into *all* of the free energy claims and *all* of the devices in person. They're *all* a hoax and free energy is not possible. But here (holds up his book)... you should buy my book. It explains how it's all possible." I said to myself... "wtf"... says it's not possible, but to read his book explaining it is indeed possible? I read it anyway. Then ended up throwing it in the trash after I was done reading and highlighting it. Didn't want that crap on my bookshelf.

A few years later, Dr. Kaku was giving an AMA (Ask Me Anything) forum on the internet. A few of my colleagues and myself were able to get him to mention something about LENR (Low Energy Nuclear Reactions) and cold fusion efforts. And again, big-headed Kaku said it aint so without doing a lick of research. I call him big-headed because he quite literally has one of the biggest heads and craniums I've ever encountered. It's Neanderthalic. Man's neck must be hurtin'. You'd figure, "the bigger the head, the bigger the brain," right? But then again, look at Homer Simpson's X ray. Michio Kaku had no clue about the updated efforts and lectures given at MIT a couple years prior to his AMA. He just had to backtrack and did his usual cop out of "I'm Michio Kaku, therefore I'm right." Just like Neil deGrasse Tyson. Think I'm lying? Look at this dome. I should've worn a shirt that said, "I'm With Stupid." Had to add the caption later.

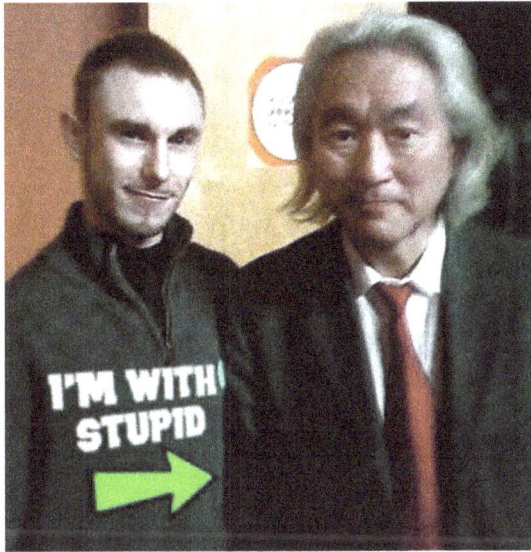

Fig. 3.8

(This guy pretty much invented String Theory trying to untangle the entire universe... but probably can't even untangle the clasps of a bra.)

There's another gentleman named Dr. Robert Koontz who shares the same feelings about Michio as I do. He's not quite fond of Dr. Kaku's rhetoric either and shares that here: *Open Letter to Professor Michio Kaku: "You Could Be Sued for Millions and Ridiculed for Centuries" (2009)*:

http://www.doctorkoontz.com/Scalar_Physics/Letters/Kaku/Open_Letter_to_Dr._Michio_Kaku.htm

I never liked String Theory and always thought it was a logical fallacy of circular thinking called tautology. "All particles are made of strings and all strings are made of particles." Never bought it. The show "Sliders" was pretty cool though. The only legitimate "string theory" I support is the work of Dr. Winston Bostick who was associated with plasma physics laboratories like Los Alamos National Laboratory, Lawrence Livermore Laboratory, Stevens Institute of Technology, MIT and Tufts University.

Nobel laureate Dr. Arthur Compton was Winston's PhD advisor. Dr. Bostick discovered plasmoids, plasma vortex phenomenon in space, plasma focus and was able to demonstrate what he believed to be observable cosmological processes, to scale. Showing that alleged Hubble expansion was not due to the deformation of space-time like how Lorentz invariance interprets it, but rather that galaxies repel each other like two electrostatic pith balls with like charge over vast distances. Dr. Bostick claimed his work with plasmas which produced filaments of particles could explain the composition of charge-carrying strings from finite bits of matter. Literally showing strings of particles in a lab but dismissed by the very people who promote string theory today as "fringe science". The disconnect is unreal.

In a talk with C-SPAN2, Kaku said, "In science, you always say, we make observations, you have a theory... you make more observations and then there's a very, very tedious process. Wrong! Nobody that I know of in my field unders—.. uses the so called scientific method. In our field, it's by the seat of your pants. It's leaps of logic. It's guesswork."
And that's "modern" science in a nutshell folks. That's the problem, and he's proud of it. Acknowledging that everyone he knows abandoned the scientific method and replaced it with "the seat of your pants", "leaps of logic" and "guesswork." These talking heads are stuck on their quantum quackery psyence of interference fringes yet they point the finger and declare *others* as fringe! Pot, meet kettle.

I'd like to have a... civilized discussion on the matter, in front of everyone. On Twitter, now X, I put forth a challenge to simultaneously fight Neil deGrasse Tyson and Bill Nye. I said, "We can make it a charity event. You can both run around the ring like charged particles until we collide. We will wear microphones and debate complex physics topics while we hurt each other real bad." I call the event "Fist Fights with Physicists."

**TheRealVerbz (Jason Verbelli)** ✔
@TheRealVerbz

I challenge you and Bill Nye to a fight. Simultaneously.

We can make it a charity event. You can both run around the ring like charged particles until we collide. We will wear microphones and debate complex physics topics while we hurt each other real bad.
Are you up for it? @BillNye

**To all those who believe that science is a consensus...
or who think, "the science is settled"...
I present you with Galileo's preserved middle finger:**

9:56 AM · Jun 20, 2024 · **1,001** Views

View post engagements

○ 4          ⇄ 4          ♡ 18          ⬜ 2          ⬆

Fig. 3.9

# Implications for Maxwell's (Heaviside's) Equations & more

Wanted to first mention about the misconception of a sine wave. When you hear a perfect 432 Hz A note tone from a tuning fork, we are hearing many sine waves connected. Not just one long wave. There are 432 new vibrations of the prongs in 1 second which then transfers that energy to the molecules making up the air and the air shakes at a frequency/rate of 432 new shakes per second. Each hertz/ vibration/ oscillation/ cycle is a new emission when it comes to EM radiation. One pulsation from an electron can be considered 1 Hertz as an EM packet. For sound and vibration of a medium, it's about shaking that medium 432 times back and forth a second. People then misapply that logic to an electron "vibrating," giving a false impression that the same dynamic happens for electromagnetic radiation. So people "oversimplify" (lie or mislead) the process using the surface of water which isn't an accurate representation at all. It's awful.

It's not just the same light or the same sound. The shaking of a tuning fork's prongs are individual reciprocations repeating at a constant. It's not just "**a** sound" or the same sound emanating from the prongs of a tuning fork. It's a brand *new* sound produced at a constant, 432 times in one second. The air is being shaken once... twice... a brand new third shake... 4 shakes... 432 individual shakes back and forth. Each shake and oscillation and vibration is ONE cycle within a series of repeating cycles in sequence... and at a constant. There are specific likelihoods of timing your actions to the "tune" of the breaks in the wave (meaning in phase or resonating).

In electrical engineering, the same premise applies but rather than dealing with a physical medium vibrating back and forth a given amount of times a second (frequency/rate), it's about absorption and emission of the energy of the electron itself in a pulsation. When we see a sinewave on an oscilloscope, we're seeing a series of brand new emissions strung together at a constant which forms the illusion of "one wave." There's a lot more to it, but I want to deliver the initial premise. Brand new emissions comprise a packet of electromagnetic radiation which sequentially repeat at temporally coherent intervals.

It's not just "a packet' or "a wave" of light. Every half cycle of the sine wave begins a new emission. Here is my logic using Galilean variance and mathematical proofs.

The conventional sine wave equation is:

$$y(t) = A\sin(\omega t + \varphi)$$

$A$ is the amplitude.

$\varphi$ is the phase shift.

$t$ is time.

$\omega$ is the angular frequency which is: $\omega = 2\pi f$

$f$ is the frequency.

The sine wave itself is broken into four quadrants. I'm using the assumption that the sine wave repeats after every half cycle. You only have to account for half of the cycle because e it's like compression and rarefaction. Adding pressure through compression automatically results in an equal and opposite rarefaction like a recoil. So the peak will initiate a trough. So the next peak is a brand new emission. New peak, new emission. A bunch of new emissions strung together at a constant is "**a** sine wave". The "**a**" in bold emphasizing people mistaking the sine wave as just being "one wave" when really it's a repetition at a constant. This is already known to electric engineers and serious people in the field.

But I'm making the distinction as it pertains to light from different frames of reference or the notion of a brand new light being produced at a constant. You break any part of that process and you're just cutting into the already segment parts of the "wave"/cycle which will initiate yet more segments being output from the electrons making up the blockage or resistance. It's not the same primary, constant, single, long wave of light distorting within the same frame of reference. Any time you block a given packet of light, it simply cuts the graph of the sine wave and pastes it to a new frame of reference. The cut and paste of sine waves results in a new frequency. And that new frequency is representative of the velocity of light itself shifting. A copy and pasted frequency shifted sine wave (re-emission) is always in phase with the primary wave that fed the electrons the initial energy. The electron gets the energy to re-emit from absorbing the energy from the primary photon packet. It's never the same elongated sinewave that stretches. There is a definitive cut-off point within the sine wave peaks and troughs that serves the ILLUSION of a smooth distortion because that cutting and pasting of sine wave parts happens at the speed of light.

People might view the sine wave like this and mistakenly put the entire peak and entire trough of the wave in its own quadrant.

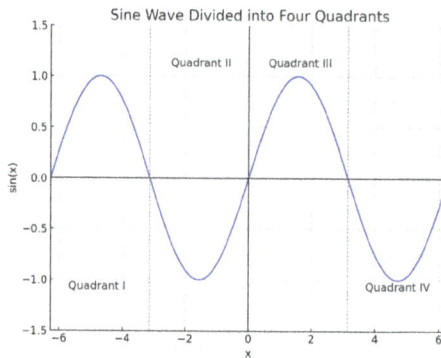

Fig. 3.10

But I break up this image into 8 sections. So we can cut off half the chart and focus on just 1 peak and 1 trough and break *that* into 4 quadrants.

Focusing on breaking up 1 full peak and 1 full trough represents 1 Hertz. 1 full cycle. A full oscillation is 1 full emission. How many emissions per second = how many Hertz.

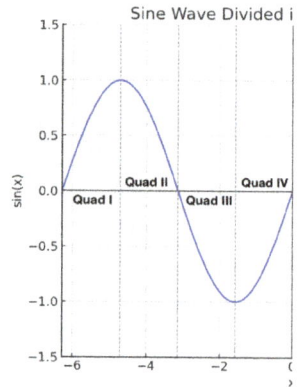

Quadrant I: First half of the peak (from 0 to $\frac{\pi}{2}$)

Quadrant I: Second half of the peak (from $\frac{\pi}{2}$ to $\pi$)

Quadrant Ill: First half of the trough (from $\pi$ to $\frac{3\pi}{2}$)

Quadrant IV: Second half of the trough (from $\frac{3\pi}{2}$ to $2\pi$)

Fig. 3.11

Next, we focus on the time it takes for each particular quadrant to take place. The "Time Interval" and "Segment Function".

For Quadrant I (The First Half of Peak):

Time Interval: $t_n = \left[ n\frac{\pi}{2\omega}, (n+1)\frac{\pi}{2\omega} \right)$

Segment Function: $y_n(t) = A\sin(\omega t + \varphi), \quad t \in t_n$

For Quadrant II (The Second Half of Peak):

Time Interval: $t_n = \left[ n\frac{\pi}{2\omega} + \frac{\pi}{2\omega}, (n+1)\frac{\pi}{2\omega} \right)$

Segment Function: : $y_n(t) = A\sin(\omega t + \varphi), \quad t \in t_n$

For Quadrant III (The First Half of Trough):

Time Interval: $t_n = \left[ n\frac{\pi}{2\omega} + \pi, (n+1)\frac{\pi}{2\omega} \right)$

Segment Function: : $y_n(t) = A\sin(\omega t + \varphi), \quad t \in t_n$

For Quadrant IV (The Second Half of Trough):

Time Interval: $t_n = \left[ n\frac{\pi}{2\omega} + \frac{3\pi}{2\omega}, (n+1)\frac{\pi}{2\omega} \right)$

Segment Function: : $y_n(t) = A\sin(\omega t + \varphi), \quad t \in t_n$

The segment functions are identical because they are talking about the same type of arc section in each quadrant, so the function of the segment is common between all quadrants. All four parts of the wave together make for 1 Hertz. Each Hertz is a brand-new Hertz. A repetition of individual Hertz is not just "a wave". Bump that up to the terahertz range and people misapply the logic of a graphical representation of a sine wave or the constant on an oscilloscope as being **a** sine wave. Or thinking of a packet of light as just one light. No... it's a brand-new light at each interval and completion of a cycle. It's never "the same light" at any point of the wavepacket. Maxwell/Heaviside's equations are ASSUMED to require a constancy in *all* frames of reference like the velocity of light. They say EM radiation and the solution to Maxwell's equations *must* mean the same energy stretches or is being exerted within the same frame of reference. But that wrongful postulate exists only under the mathematical framework of Lorentz invariance. But we are dealing with Galilean variance. Making serious distinctions that relativity and invariant models does not.

(I used Pi for the sine wave equations because it's easier to communicate for now. But in reality, Tau should be used, ala Michael Hartl https://tauday.com/tau-manifesto )

James Clerk Maxwell originated the equations for electric and magnetic fields. There were 20 different equations with 20 different variables. Oliver Heaviside then came along and reformulated Maxwell's equations. Heaviside doesn't get enough credit. People refer to "Maxwell equations" in regard to electromagnetism, but it was Heaviside who consolidated/ truncated the 20 equations down to 4 using vector calculus. Michael Faraday discovered electromagnetic induction which then led to Faraday's Law.

Unto himself, André-Marie Ampère discovered that an electric current produces a magnetic field. Johann Carl Friedrich Gauss made significant contributions to the scientific community in regard to optics, gravity, electromagnetism, astronomy and more. All of these guys made huge contributions. Maxwell refined their works, and then Heaviside along with Josiah Willard Gibbs who reformulated Maxwell's corrections from there.

Using Dr. Dowdye's *Extinction Shift Principle*, we can perhaps refine a bit more. We can derive new equations under the assumption that the velocity of light shifts in different frames of reference, and then re-emits from the electrons making up the interference. A *variance* under Galilean transformations reformulated by Edward Dowdye. Under this model:

- There is *no* particle-wave duality. (Mass and energy do not literally interchange.)
- The speed of light changes. (The velocity of light is variable.)
- Quantum Theory and Quantum Electrodynamics (QED) relies on photons being particles and a Lorentzian invariance of light speed. (Requires the velocity of light be the same to everyone *no matter what* in *all* frames of reference.)

- This model replaces the Lorentz invariance with a Galilean variance.

The assumption that Maxwell/Heaviside's equations require a constancy in ALL frames of reference is invalid and relies on a relativistic/ Lorentzian invariant model for light. The consequences of this figuratively ripples throughout the scientific community like an earthquake's P, S and L waves. This affects the foundation of quantum theory and more.

As Oliver Heaviside himself wrote in a letter to Vilhelm Bjerknes, dated March 8th, 1920, "I don't find Einstein's Relativity agrees with me. It is the most unnatural and difficult to understand way of representing facts that could be thought of. And I really think that Einstein is a practical joker, pulling the legs of his enthusiastic followers, more Einsteinisch than he."

Conventional Faraday's Law of induction:

$$\nabla \times \mathbf{E} = -\frac{\partial \mathbf{B}}{\partial t}$$

Modified Faraday's Law where the propagation of $V$ shifts with frames of reference:

$$\nabla \times \mathbf{E} = -\frac{1}{v}\frac{\partial \mathbf{B}}{\partial t}$$

Conventional Ampère's Law:

$$\nabla \times \mathbf{B} = \mu_0 \mathbf{J} + \mu_0 \epsilon_0 \frac{\partial \mathbf{E}}{\partial t}$$

Modified Ampère's Law with Maxwell's correction under Galilean variance:

$$\nabla \times \mathbf{B} = \mu_0' \mathbf{J} + \mu_0' \epsilon_0' \left( \frac{1}{v}\frac{\partial \mathbf{E}}{\partial t} \right)$$

Regarding the so called permittivity and permeability of space...
I think space is another word for "distance." The space between planets, the distance between planets. The void of space describes the vastness of the cosmos. There are particles here and there in space, and there are massless WAVEPACKETS that pervade all of the void. But the misconception is in a non-existent particle-wave duality or the belief that

mass and energy literally interchange. Which means the assumption of Dirac's sea of energy means there must be a probability of a particle to pop up at any given point based upon the idea of quantum states and superposition, etc. So there are particles here and there.. but what's in between those particles? Even the model of an atom says it's 99% "empty space" *by volume.* And since they insist matter and energy are literally interchangeable... and there's energy between planets even though there aren't particles everywhere between planets... the idea is that energy can convert at any given point in space; therefore, they say all of "space is not empty."... but it mainly is.

The scientific community has an extreme aversion to emptiness and the concept of 0 or no medium between planets and atoms. People get so pissed off when you suggest empty space or the concept of zero, they foam at the mouth like a rabid relativist.

Try it out for yourself. Be cordial, calm and articulate and then watch the ad homonyms fly. Works every time. Great for stretching the schadenfreude muscles. I aim for hypertrophy of my schadenfreude muscles. My laughter reverberates through time and space. And the rate at which my laughter can propagate through space is dependent upon the permittivity, $\epsilon_0$ and permeability, $\mu_0$ of the void. (distance/space can't have permittivity or permeability.) Space is a measurement of distance in a given increment. Distance/space doesn't curve or bend or warp or ripple or distort or flatten or expand or contract, etc. Distance is a concept, not something concrete that has physical attributes. Trying to turn a concept into something concrete that behaves like condensed matter is a logical fallacy called "reification." And since there is no physical space-time or any medium at all... and all electromagnetic radiation is massless, there can be no permeability or permittivity of a concept. The variables still apply but need to be reworded as to what they apply to. Semantics matters, so the equations are the same, but modified for what they are applied to. Thus the apostrophe/prime notation above the epsilon' and mu' and c' symbols.

$$\epsilon_0' = \frac{1}{c'^2 \mu_0'} \quad \text{and} \quad \mu_0' = \frac{1}{c'^2 \epsilon_0'}$$

So, graphically, there are definitive lines of demarcation between emissions and it's not just one wave stretched within the same frame of reference due to distortions of space-time...

The conventional Time-Dependent Schrödinger Equation:

$$i\hbar \frac{\partial \psi}{\partial t} = -\frac{\hbar^2}{2m}\nabla^2\psi + V\psi$$

The Modified Time-Dependent Schrödinger Equation:

$$i\hbar \frac{\partial \psi}{\partial t} = -\frac{\hbar^2}{2m}\left(\frac{1}{v^2}\right)\nabla^2\psi + V\psi$$

The conventional Klein-Gordon Equation:

$$\left(\frac{1}{c^2}\frac{\partial^2}{\partial t^2} - \nabla^2 + \frac{m^2v^2}{\hbar^2}\right)\phi = 0$$

Modified Klein-Gordon Equation with variable velocity $v^2$ instead of $c^2$:

$$\left(\frac{1}{v^2}\frac{\partial^2}{\partial t^2} - \nabla^2 + \frac{m^2v^2}{\hbar^2}\right)\phi = 0$$

Conventional Dirac Equation:

$$\left(i\hbar\gamma^\mu\partial_\mu - mc\right)\psi = 0$$

Modified Dirac Equation under Galilean variance:

$$\left(i\hbar\gamma^0\frac{1}{v}\partial_t + i\hbar\gamma^i\partial_i - mv\right)\psi = 0$$

Conventional Heisenberg's Commutator Relation:

$$[x,p] = i\hbar$$

Modified Commutator Relation under Galilean variance:

$$[x,p] = i\hbar\left(1 + \frac{v}{c}\right) \quad \text{or} \quad [x,p] = i\hbar f(v)$$

Conventional Heisenberg Uncertainty Principle:

$$\Delta x \Delta p \geq \frac{\hbar}{2}$$

Modified Heisenberg Uncertainty Principle under Galilean variance:

$$\Delta x \Delta p \approx \frac{\hbar}{2}\left(1 + \frac{v}{c}\right)$$

Classical physics was abandoned originally because it used Galilean invariance. Relativity was invented using Lorentz invariance. Dr. Edward Dowdye's Extinction Shift Principle (ESP) reformulated classical physics equations so that Galilean transformations are now variable. Below are a few side-by-side comparisons.

### Side-by-Side Comparison with Equations

| Aspect | Quantum Interpretation | Classical Interpretation (Dowdye) |
|---|---|---|
| Wave Equation | $\nabla^2 \psi - \frac{1}{v^2}\frac{\partial^2 \psi}{\partial t^2} = 0$ | $\frac{\partial^2 \psi}{\partial r^2} + \frac{1}{r}\frac{\partial \psi}{\partial r} + \frac{\partial^2 \psi}{\partial z^2} - \frac{1}{v^2}\frac{\partial^2 \psi}{\partial t^2} = 0$ |
| Wavefunction Superposition | $\Psi = \psi_A + \psi_B$ | $\Psi = \sum_{i=1}^{4} A_i \cos(kr_i - \omega t + \phi_i)$ |
| Interference Pattern | $I$ | $\Psi$ |
| Energy Description | Schrödinger Equation: $i\hbar\frac{\partial \psi}{\partial t} = -\frac{\hbar^2}{2m}\nabla^2 \psi + V\psi$ | Hamiltonian: $H = \int\left(\frac{1}{2}\epsilon_0\left(\frac{\partial \Psi}{\partial t}\right)^2 + \frac{1}{2}\frac{1}{\mu_0}(\nabla\Psi)^2\right)dV$ |
| Modified Parameters (ESP) | Not applicable | $k' = k\left(1 + \frac{v}{c}\right), \omega' = \omega\left(1 + \frac{v}{c}\right)$ |
| Re-emission Boundary Condition | Not applicable | $\psi(r,t) = A\cos(kr - \omega t + \phi)$ |

Fig. 3.12

In order to produce interference fringes using particles, you need a very specific setup to manipulate the parameters. They pass an electric current through a filament made of tungsten, and it glows red hot like a light bulb. So, it's just a tungsten light bulb. It produces a spherical wavefront because it's light from the bulb which is out of phase. They are not concerned about the light through the slits in this experiment though. They are only concerned about the dots forming from the emissions of electrons shooting off from the tungsten in that experiment. In the box housing the 4000°F filament, there would be

2 bands of light on the screen which we can't see because it's inside a closed box. But the detector screen gets impacted by the particles over time and they produce a multitude of interference fringes. That's what that experiment aims to capture. The formation of patterns from ejected particles. The spherical wavefronts of the light produce 2 strips of light, but the particle emissions show interference fringes at the same time inside the box. The misnomer is that the particles and the waves are one in the same. There is no differentiation between the electron emissions verses the light in quantum theory. They say the patterns line up, yet they are separate in the actual experiment, so that *must* mean a particle-wave duality. The light forms 2 bands but the particles form many. Electrons produce plane wave interference and the light produces 2 strips of light.

How does one create a thermionic emission of electrons to produce just 2 bands on the screen? That would be achieved without use of high voltage or focusing with a magnetic lens. That would produce a spherical wavefront analog. Sure, you also get secondary emissions of electrons but there is no plane wave present to hit the slits with a definitive probability in a wave-like diffraction pattern. All reflection, refraction, transmission, diffraction, etc. are *all* re-emissions. There's just a major difference in interpretations.

 Thermionic emissions of electrons assisted by high voltage and electromagnetic lenses can mimic a plane wave to produce multiple cylindrical bands of light and dark on a detector screen after the electrons pass through slits. The absence of high voltage or an electromagnetic lens is analogous to a spherical wavefront.

The double slit experiment becomes the single slit experiment when they "measure" because they physically block one of the slits with a sensor to prevent the formation of the particle pattern. They depict the scenario in cartoons as if observation alone IS measurement. As if a real-world human is looking at the unchanging slits. But they change the factor of the scenario by blocking one of the slits with the sensor itself in the particle experiment. There are different experiments, all of which produce fringes or not, but not because of some mythical particle-wave duality.

Light can be sent through two different ways to produce two different patterns. That does NOT mean the light and particles magically interchange, or literally swap or that there is ANY literal duality at all. Sand can behave LIKE liquid during an earthquake like liquefaction. But the sand didn't literally BECOME a liquid or water. It just flowed LIKE a liquid. Not that there is a literal interchange or phase transition of the sand particles to a wave of energy and then back again. There seems to be less discussion about the difference in patterns produced by plane waves and spherical waves for BOTH particles and waves... and too much nonsense science fiction talk about non-existent particle-wave dualities. I'd like to have a word with your science manager please, I'd like a refund.

So, the coherent set up produces a wave-LIKE pattern for the particles and an actual wave pattern for coherent light. OR... produces an actual wave pattern from a plane wave of incoherent light or *any* type of light as long as it's a plane wavefront. But it's still an assumption that the thermionic emission is a one-by-one single file line of electrons. And that those electrons from the filament are the same electrons hitting the detector screen. A primary electron shooting out from the filament just might hit the edge of the slit and induce a secondary electron emission from the slit itself. Which would contribute to the pattern as well. In addition to the probability of the electrons forming multiple patterns on the wall which obviously increases and is demonstrated because there is a definitive coherent distribution of particles in an undulation from the artificially induced lens set up. So, there is no particle-wave duality. There is no delayed choice. Reality is not subjectively manifested, and a lot of quantum electrodynamics is pretty much erroneous.

This is beginning to look like an allegory of Plato's quantum cave. The shadows, and the puppets making the shadows are two different things. The puppets and the shadows do not interchange or magically flip back and forth. The puppets and shadows are not in a probable state of quantum superposition, etc. When you block the light, you get a shadow.

Just imagine not being able to discern the difference between the factors of a scenario. You only get an interference pattern with particles if you don't block the beam. The 2 slits must fit within the diameter of that beam. If you block one of the slits or prevent the secondary particles from making it to the wall... you're not gonna get an interference pattern. What does that possibly have to do with some mythical particle-wave duality where a LITERAL flipping is claimed to take place? The "simplification" in textbooks is based upon a real-world observer with human eyes looking at the wall. There is a major difference between "OBSERVING" light and observing a scenario from afar... versus DIRECTLY BLOCKING a collimated beam with a sensor to prevent a pattern from forming. Just because you look at the wall doesn't mean the pattern will change. But that's how the double slit experiment is framed in the cartoon called "Flatland" with "Dr. Quantum". It's framed that way in textbooks depicting a human eye literally looking at the scenario and more. When I have discussions with people about it, they had no clue about the real setup.

At what point does "oversimplification" just become flat out fraud? How far does a misrepresentation have to go to become an all-out lie or fabrication? 100 years of the world's leading experts and universities, cartoons, diagrams, books and people showing you an incorrect model of the prism, light, darkness and color. People trust the same organizations who have failed them for decades. Whether it be science, politics, medicine, or any aspect of life really.

People are still allowed to think the double slit experiment involves literal human observation but it doesn't. This can't be. The gross negligence is too great to bear. It MUST be purposeful deception at this point. I'm having difficulty believing that generations of people can be this insane and never even ask relatively simple questions for that level of so-called "expertise."

The double slit experiment producing particle patterns has a lightbulb with a tungsten filament. The filament glows red hot at about 4000°F. That hot source is only millimeters or even inches away from the "wall" with the slits. (All of which is inside a small box on a tabletop.) The tungsten gets so hot that it shoots off electron particles like a nuclear material shooting off alpha and beta particles. The particles wildly shoot around, but the particles are charged. Which means you can control them and guide them into a collimated beam with magnetic fields. But there's no accounting for the edges of the slits themselves being impacted by the electrons from the primary source and then re-emitting secondary electron emissions. There doesn't seem to be any differentiation between the primary electrons in a beam shooting through the slits verses secondary electrons shooting off or deflecting from the edges of the slits themselves. And if you simply block one of the slits with a detector or sensor... then of course you're not going to get an interference pattern like before because you're blocking the damn particles.

That has absolutely nothing to do with some alleged particle wave duality, delayed choice, quantum eraser effect nor does it suggest or support ANY notion of a literal particle wave interchange as $E = mc^2$ is commonly interpreted. There is an equivalence... not a literal interchange. I am at a loss to understand the logic, or lack thereof, pertaining to some mythical back and forth flipping of particles to waves because someone "observed" the scene or not.

The entire scenario and explanations and depictions are lies. Not just an "oversimplification." To claim that quantum theory is so complex that no one can understand it is a cop out. Yet we must believe the same people who started QED under the wrongful assumption that light speed is ALWAYS the same no matter what? Even though these same originators admitted they didn't know wtf they were talking about either and chalking every fallacy and paradox off as "a quantum mystery." What a convenient cop out to continue riding the gravy train of quantum quackery, raking in big grant money and misallocating tax money. Must be nice.

So why should anyone continue to listen to these freaks? Because... math? Ok... well now that we have Dr. Dowdye's Extinction Shift Principle and a legitimate alternative under Galilean variance. There is full justification and zero excuse not to return to classical interpretations again and without any paradoxes. While abiding by the laws of thermodynamics and the rules of Dr. Robitaille regarding intensive and extensive properties. We can now derive equations to rid the esoteric nonsense from the scientific community based from real fundamentals and natural laws.

The more you look into the history of the theories and reality of the experiments and data, the more horrified you get. You realize Lorentz invariant transformations screwed up the scientific community to such an astronomical magnitude. The current dogmatic establishment keeps dead theories propped up like a politician or like *Weekend at Bernie's.* People continue even though the Theory of Relativity isn't able to stand on its own legs or speak for itself anymore. A theory could be tragically outdated and erroneous yet many people will *still* support it. Look at flat-earthers, aether theorists and relativists. One in the same. People can't seem to let it go. Be like the Buddha... practice detachment.

Dismissing the validity of the Galilean variance of light and the work of Dr. Edward Dowdye is like a flat earther dismissing geodesy. Relativists are even worse though, because they think space is flat! I like asking relativists, "Isn't that right flat-spacer? To think space-time curves in the presence of a 3D mass must mean you think space-time is flat in between those dense masses... which is the excuse for why light propagates in a straight line in relativity's model... right flat-spacer?"

Imagine being able to yield the same predictions and solutions as relativity and quantum theory but in a simpler manner using only elementary math, 3D Euclidean Geometry, and classical physics. No extraneous relativistic procedures or corrections are needed.
Why in God's name would you want to continue using a convoluted mathematical route to get to the same answer when there is an easier alternative?! Especially when most of relativity is impossible to even visualize. (How convenient) "Space-time is one helluva drug!"

If you can't visualize something in your mind's eye... it might not be because your brain isn't capable. It might be because the scenario you are trying to visualize is bullshit. Which might give a person a false impression they suffer from aphantasia or that they should defer their critical thinking to those who seem like "they got this." All while the real sufferers of aphantasia pawn of models that can't be visualized to everyone else. They are the ones who substitute their own lack of mind's eye visualizations for computer simulations. All while claiming THEY are the ones with hyperphantasia and that we should trust their cartoons. Yikes... G.I.G.O. (Garbage In, Garbage Out)

What the scientific community has done for the last century is incredibly, ingeniously deceitful. The same empirical evidence and data can be explained under Galilean variance without need for a veil of illusions based upon Lorentz invariance. Everything from the Hughes-Drever experiment to Michaelson-Morley's interferometer to GPS, atomic clocks, the double slit, Hafele-Keating, the Sagnac effect, Ives-Stillwell experiment, etc. are all misinterpreted under the non-Euclidean 4D framework of Lorentz.

No more need for a relativistic cult of quantum quackery under Lorentz invariance.

Fig. 3.13

Fig. 3.14

— CHAPTER 4 —

# LASER Beam on a Spinning Mirror Gedankenexperiment

Albert Einstein was a deep thinker. He enjoyed thought experiments he called "gedankenexperiments." Some things we can only speculate about and play around with ideas because testing those ideas are beyond technological capabilities. You are subject to the consequences of the model you use. Einstein indeed stayed true to Lorentz invariance. Unwavering dedication to an egregiously erroneous model, but the man indeed had confidence in the framework he stood upon. Towards the end of his life, however he indicated that he questioned his own theories. Einstein seemed concerned that the scientific community might be too attached to relativistic interpretations. But it was too late, Einstein started the momentum of the beast and it took until 1991 to expose a catastrophic break in the tracks. The claims of relativity are nuts, but those are the consequences of Lorentz invariance.

"My opinion about Dayton Miller's experiments is the following. ... Should the positive result be confirmed, then the special theory of relativity and with it the general theory of relativity, in its current form, would be invalid."
Albert Einstein, 1925 (Einy was talking about Dayton Miller's "aether drift" experiments which was similar to the Michaelson-Morley interferometer and Hammar experiments.)

"I believe that I have really found the relationship between gravitation and electricity, assuming that the Miller experiments are based on a fundamental error. Otherwise, the whole relativity theory collapses like a house of cards."
Albert Einstein, 1921 (Again, Uncle Al was talking about any confirmation that the velocity of light shifted. Whether through an all pervasive static or dynamic aether around Earth.)

"You imagine that I look back on my life's work with calm satisfaction. But from nearby it looks quite different. There is not a single concept of which I am convinced that it will stand firm, and I feel uncertain whether I am in general on the right track.

Albert Einstein, on his 70th birthday, 1949 ("One-Stone" demonstrating healthy skepticism.) But as Nikola Tesla said, "The scientists of today think deeply instead of clearly. One must be sane to think clearly, but one can think deeply and be quite insane."

We will be thinking clearly under the framework of Galilean variance for this gedankenexperiment.

When you point a laser beam at a mirror, it's not the same light bouncing back. Light does not ricochet. It is absorbed and re-emitted. But what if the mirror is rotating on its axis? (Spinning like a basketball on your finger in place.) The electrons making up the mirror wouldn't be able to reflect the beam exactly 180 degrees compared to the source it came from. The electrons making up that part of the mirror would shift their position by the time they re-emit and the light would "reflect" and "drag" at a tangent. If the mirror spun fast enough and the beam was offset from the center enough, the beam that would normally reflect as a singular beam would instead reflect as a hollow cylinder of light (a ring).

The spinning mirror would generate re-emissions in different frames of reference relative to the position of the laser pointer/source. That means the light would be redshifted slightly and you would see a slight gradient of a hue around the ring. And that gradient would rotate at different rates depending on the centripetal spin of the mirror and the offset of the beam from center. The cylinder of light would be like a superposition of the singular beam, but re-emitted from a different frame of reference, at a constant.

In the thought experiment, I had to account for the reality of the tensile strength of the glass under centrifugal forces because you can't spin a mirror at relativistic speeds without having it explode apart after a critical threshold. If you had a circular mirror that was about 26.4 miles in diameter, the maximum it could spin is 1350 RPM before reaching the threshold of failure from spin forces. And you'd need a mirror that size to generate a ring of re-emitted light which would redshift in a gradient. And the beam would need to be offset from the center of the spinning mirror by about 13 miles/ about 69,500 feet.

You would see the laser beam form a dot on the surface of the mirror with a normal reflected beam at the center of the mirror. But after a certain point when you offset from the center, a ring would form. And that ring would be a superposition of reflections of the primary laser beam, but in a cylinder/ring being re-emitted at a constant from different frames of reference simultaneously. Because the re-emissions happen at the rate of c. So if you spin fast enough, the electrons don't get a chance to re-emit before they change position on their axis, thereby re-emitting a cumulative cylinder of reflected beams, AS IF the beam "frame dragged" or was "dilated" or "redshifted", "stretched," etc.

This is not the same as a Lissajous pattern because the mirror is not vibrating, but rather would be rotating without vibration. Special Relativity doesn't distinguish between a linear anomaly versu s a rotational anomaly. Lorentz invariance assumes special relativity takes effect in rotating systems like they do with a linear accelerating body.

If you ACCELERATE the mirror up to a high speed and point the beam at it... relativity says you'd get a phenomenon after a certain RPM like I'm describing, but only while accelerating. But if the mirror is just spinning at light speed in place and you point a beam at it, under Lorentz invariance there should be no effect because it's not accelerating. Just a different in spin rates between you and the rotating mirror.  Under Galilean variance though, our gedankenexperiment will indeed produce an anomaly due to the re-emissions from different frames of reference. In a rotating system, what is the system accelerating relative to? That would be different points offset to the center relative to the stationary observer. Like a Faraday disk or Bruce DePalma's experiments with rotating copper disks. The spinning disks produce magnetic fields and things like that. You get an effect, period. Not just from acceleration, but from a difference in relative spin.

# Old Shift, New Shift - RedShift, No BlueShift

Another point I want to make is that the alleged expansion of the universe is said to be verified because of redshift and Doppler shift. But red shifting does not depend on moving towards or away from an observer. ANY change in frame of reference or relative velocity will be ALYWAYS be viewed as a redshift. Blueshift is not possible in Galilean variance or the Extinction Shift Principle. Therefore, the entire concept of universal expansion is bogus and erroneous, just for that discrepancy of linear acceleration and rotational velocity.

I began putting together an interactive page with slide controls and toggles to mess around with the idea and then played with the math a bit.

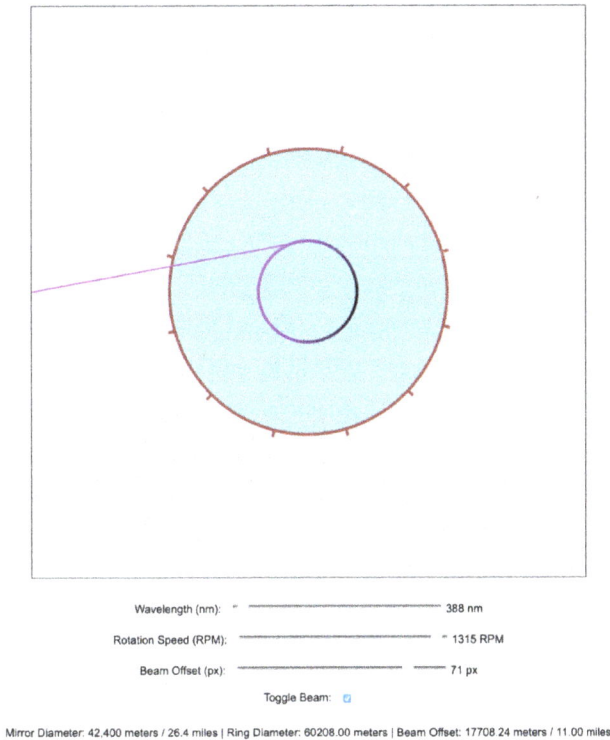

| | |
|---|---|
| Wavelength (nm): | 388 nm |
| Rotation Speed (RPM): | 1315 RPM |
| Beam Offset (px): | 71 px |
| Toggle Beam: | ☑ |

Mirror Diameter: 42,400 meters / 26.4 miles | Ring Diameter: 60208.00 meters | Beam Offset: 17708.24 meters / 11.00 miles

Fig. 4.1

| Parameter | Metric Units | Imperial Units | Imperial Units (miles) | RPM |
|---|---|---|---|---|
| Desired Ring Radius (R) | 0.01 meters | 0.033 feet | 0.0000062 miles | - |
| Distance to Screen (D) | 1 meter | 3.28 feet | 0.000621 miles | - |
| Low Rotational Speed (ω) | 10 rad/s | 10 rad/s | 10 rad/s | 95.5 RPM |
| High Rotational Speed (ω) | 1000 rad/s | 1000 rad/s | 1000 rad/s | 9550 RPM |
| Offset from Center (r) at Low ω | 300,000 meters | 984,252 feet | 186.4 miles | - |
| Offset from Center (r) at High ω | 3,000 meters | 9,843 feet | 1.864 miles | - |
| Diameter of Mirror (d) at Low ω | 600,000 meters | 1,968,504 feet | 372.8 miles | - |
| Diameter of Mirror (d) at High ω | 6,000 meters | 19,686 feet | 3.73 miles | - |

Fig. 4.2

Now let's take this to the next level using two counter-rotating mirrors facing each other. So the incoming laser beam hits one of the mirrors at an angle of incidence of 45° which bounces (re-emits) the beam between the mirrors as they counter-rotate.

That would produce a greater redshift effect at a lower RPM with smaller diameter mirrors. Assuming the mirrors could each spin 10,000 RPM, with a 20,000 RPM relative difference in velocity between them, you'd "only" need mirrors with a diameter of 572 meters. The redshift effect of the laser light would take place at a radius of 286.48 meters offset from the center of the counter-rotating mirrors. Any re-emitted/reflected light less than that radius would not produce the redshift effect or make a ring/cylinder. But any parts of the beam re-emitting/ reflecting from the surface greater than the radius of 286.48 m would produce the effect.

| Parameter | Value |
| --- | --- |
| Mirror Diameter | 572.96 meters |
| Radius for Redshift Effect | 286.48 meters (offset from center) |
| Required Rotational Speed | 10,000 RPM (for each mirror) |
| Required Thickness | 3.56 µm |
| Redshift Effect Details < | Light reflecting between the two mirrors less than the radius of 286.48 meters would not redshift. |
| Redshift Effect Details > | Light reflecting between the two mirrors greater than the radius of 286.48 meters would redshift. |

Fig. 4.3

You can play around and increase the RPM which will require a lesser diameter mirror and radius offset from center. You sacrifice one variable to adjust the others to maximize the feasibility of the experiment. But also must remain within the reality of physical forces.

Again, according to relativity and Lorentz invariance, the red shifting should not take place since there is no acceleration between the two mirrors. Only a static relative difference in velocities. Relativity doesn't discriminate between linear acceleration and rotational velocity so that would be yet another violation of relativity to demonstrate.

But of course, relativists would invent some new equation or new force or new particle under Lorentz invariance as an excuse for the violation of their model. "It's because of… uh… dark matter… yeah that's it!" Or "it's because of a manifestation of virtual pair production from the 4th dimension emerging as a pseudoforce of super-strangeness and….." Dude… just shut up… These people sound like that guy in the skit talking about the "turbo encabulator." They just ramble off a bunch of obfuscating nonsense, like a sophomaniac using sesquipedalian words as if they were an ultracrepaderian. (See… I can do it too!) Translation: Inventing concepts to keep your Lorentz invariant model from collapsing is the epitome of the Dunning-Krueger Syndrome on steroids. And that's relativity for ya!

Wanted to briefly mention the assumptions relativity makes about the Andromeda galaxy and a couple of other fallacies about alleged distorting of light.

Red shift does not depend on acceleration or direction. Relativists think redshift is only seen from bodies accelerating away from us. But redshift occurs with any difference in relative velocity regardless of direction. No need for acceleration.

So if a galaxy is moving toward us, but not accelerating, it will redshift.
If a galaxy is moving towards us and is accelerating, it will redshift.
If a galaxy is moving away from us, and is accelerating, it will redshift.
If a galaxy is moving away from us, but not accelerating, it will redshift.

Any difference in relative velocity results in redshift. Redshift is not intrinsic or based upon the age of a galaxy or the factors Halton Arp speculated. Halton Arp cataloged peculiar galaxies, meaning ones that didn't toe the line for relativity. Galaxies that exhibit no redshift or a different redshift than the galaxies they are connected to through luminous bridges called Birkeland currents. Birkeland currents can stretch thousands of lightyears or much more. Some galaxies thought to be connected were distant from each other while other galaxies thought to be distant were actually connected. The redshift of one galaxy should match another if they are connected according to relativity, but observation concludes they do not redshift like how relativity predicts. It's yet another direct violation.

After Halton Arp published his books showing evidence that contradicted the Big Bang model... the astrophysics community banned Arp from observatories and refused to publish his research. Journals refused to publish or review his papers. Rather than say, "wow, thanks Halton, here's your Nobel prize," the establishment tried sweeping his findings under the rug. That is not gross negligence and innocent oversight. That's dogmatic fraud to keep a dead model seem like it's still alive like Weekend at Bernie's. Halton Arp was treated no different than how the Vatican treated people hundreds of years ago for being heretics. This was explained in an article from 11/01/2004 titled, "*Halton Arp: A Modern Day Galileo.*"

There were galaxies that didn't redshift in the zones where other galaxies did display redshift. That put a wrench in the total theory since the values for space-time expansion was supposed to be of even values for galaxies the same distance away from Earth.  But they aren't the same for the "peculiar galaxies" Halton Arp observed and documented.

But Arp mistook the *cause* of redshift. The damage was done to his career though. The exposure of the peculiar galaxies in his books itself was enough to get Arp blackballed. His books are called, "*Quasars, Redshifts and Controversies*", "*Seeing Red*," and "*Catalogue of Discordant Redshift Associations*."

The other thing I wanted to mention was about the non-existence of blueshift. Blueshift does not exist because ANY difference in relative velocity results in a redshift. It's not dependent upon direction, like accelerating away gets you redshift and accelerating towards you gets you blue shift... no. Doesn't' matter towards or away, accelerating or not. It *all* results in a redshift of light relative to a stationary observer.

The best "evidence" for blueshift is claimed for the Andromeda galaxy. And Andromeda isn't even blue shifting... it's just NOT red shifting. Because there is no redshift of Andromeda, it's automatically *assumed* Andromeda *must* be moving towards us. And if it's moving towards us... it will accelerate... and if it accelerates then it will *eventually* blueshift. It's pure assumptions based on an invalid and invariant model for light.

| Factor | Lorentz Invariance | Galilean Variance |
|---|---|---|
| Nature of Time | Time is relative and dependent on the observer's frame. | Time is absolute and the same for all observers. |
| Speed of Light | Constant in all inertial frames, independent of the source or observer's motion. | Variable depending on the relative motion between source and observer. |
| Transformation Equations | Uses Lorentz transformations, which account for time dilation and length contraction. | Uses Galilean transformations, where time and space are independent, and there is no length contraction or time dilation. |
| Inertial Frames of Reference | All inertial frames are equivalent, but time and space are intertwined, leading to different measurements of time and space in different frames. | All inertial frames are equivalent, but time and space are treated separately, with time being absolute. |
| Redshift and Blueshift | Both redshift and blueshift are possible, depending on the relative motion. | Only redshift occurs due to the extinction of light, with no blueshift, irrespective of direction or acceleration. |
| Space-Time | Space and time are unified into a single four-dimensional continuum (space-time). | Space and time are distinct entities, with no concept of space-time curvature. |
| Gravity's Role in Inertia | Gravity is described as the curvature of space-time, influencing inertial frames. | Gravity is not linked to space-time curvature but is considered a force like any other in classical mechanics. |

Fig. 4.4

# William de Sitter's Double Star Misnomer

The same light can never distort. A brand-new light is re-emitted from the point of extinction. What is an "extinction" in astrophysics? An extinction is the absorption and the scattering of electromagnetic radiation by interstellar and intergalactic media like dust, grains or gas clouds. This is a well-accepted phenomenon, but it seems the logic of the scenario escapes relativists. Extinction was first proposed around 1915 and 1916 by Paul Peter Ewald and Carl Wilhelm Oseen. A man named J.G. Fox, who worked on the Manhattan Project, misapplied the concept of Extinction in the 1960's in an attempt to substantiate Special Relativity using Lorentz invariance. Fox, Walther Ritz, Wolfgang Pauli and other emissions theorists were on the right track but were stifled due to the introduction of the particle-wave duality fallacy. So the ballistic principles went dormant until Dr. Dowdye's Extinction Shift Principle correctly applied Galilean transformations but under a variable framework now instead.

The light from a distant star or source will have to pass through layers of media before making it to our solar system and then reach an observer on Earth. A similar application of the effect is seen in images peddled by NASA and touted as evidence of gravitational lensing. The intermediary dust and "stuff" that's present will absorb any incoming light and will then re-emit a new pack from the frame of reference of the dust and "stuff."

When the light experiences an extinction and is then shifted to a new frequency, that is an extinction shift. Thus, Dowdye's ESP/ Extinction Shift Principle reiterated. But that applies to *all* electromagnetic radiation *and* emissions of gravity. And the interfering media is the electrons making up all matter and observers themselves. The velocity of any packet of light will be extinction shifted as a new secondary packet from a new frame of reference. Not that a primary packet distorts in the same frame of reference due to deformations in space-time.

The entire basis of Schwarzschild geodesics is thereby erroneous since that approach derives its conclusions from the assumption that the velocity of light is invariant. Examples claimed to be some of the best validations of relativity are some of the predictions for the perihelion of mercury, Wilhem de Sitter's double star experiment, and binary pulsar system 1913 and 1916. Supporters of relativity seem to still believe that Lorentz invariance and relativistic procedures are the only possible method to arrive at the correct predictions for the aforementioned scenarios. Sure, you can use Schwarzschild geodesics AS IF it's valid because it is indeed applicable to accurate predictions. But it's not an accurate depiction of how reality actually arrives to the data obtained. Gravity is not some mythical curvature of space-time.

In Figure 11A on page 29 of Dr. Dowdye's *Extinction Shift Principle* book, it begins the detailed dissection of the perihelion of Mercury with a few pages of corrections. Dowdye also goes into detail on the binary pulsar scenario on the pages following the dismantling of the perihelion of Mercury under Galilean variance.

Here is an apropos quote from Dr. Dowdye's website ExtinctionShift.com. "The Willem de Sitter double star argument is easily refuted due to the countless findings in Astrophysics and astronomy pertaining to the interstellar and intergalactic media. It is clearly seen that the primary emissions from the direct sources of emissions stand a ghost of a chance reaching the Earth based observers, as the primary emissions are immediately extinguished by the predominant unseen matter, the gases and 'secondary sources' of the interstellar and intergalactic media." The statement is on the FAQ page about why we should question relativity. There are three sections of questions and answers. The quote can be found at the last portion of "Answers I".

https://web.archive.org/web/20220105083912/http://extinctionshift.com/FAQ.htm

Any dust or gas clouds become a refracting medium with an index of refraction of $n(v_o)$ given that medium is fixed in relation to the distant observer. That is from the frame of reference of the fixed medium. If the medium is moving, the index of refraction is $n(v)$.

Detailed diagrams can be found in Dowdye's Extinction Shift Principle book on pages 20A through 22A. Here is my own crude diagram regarding the double star misnomer:

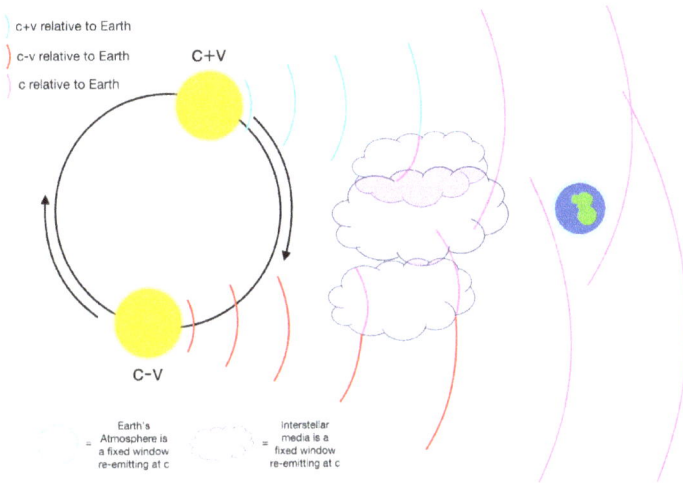

William de Sitter's double star experiment is easily refuted given that it's never the same light making it to Earth from any distant source. The electrons making up interstellar media absorbs any incoming light from the two stars, and those same electrons re-emit a brand new light which shifts frequencies. The shift in frequencies IS a change in the velocity of light from the interfering medium's own frame of reference. Relativity forbids the velocity of light from changing so de Sitter wrongfully attributed the shift in frequencies to space-time distortions under Lorentz invariance. Even if there was no interfering media between the double stars and Earth, the Earth's own atmosphere still acts as a fixed window re-emitting all light from its own frame of reference. Therefore ALL light will ALWAYS be measured as c. Serving the illusion relativity is valid.

Fig. 4.5

It doesn't matter if they are two stars, two laser pointers, two flashlights, etc. The emissions from ANY distant sources will be absorbed and re-emitted from a new frame of reference long before any light could possibly be measured from our own frame of reference here on Earth. Even in a purely hypothetical scenario where space is totally devoid of matter between the two stars and a measuring device, the electrons making up the measuring device itself will absorb the incoming light from both stars. The electrons making up the measuring device will re-emit a brand-new secondary packet. That new secondary travels relative to the electrons making up the measuring device. All measuring devices are only capable of measuring the frequency of interference between the distant sources and the electrons making up themselves at the velocity of c. All attempted measurements will yield a nullified result or will yield the result as always being c. It's an illusion.

I present the same argument with a single laser pointer fixed window of a glass of water: Velocity of Light - Extinction Shift Example:

https://rumble.com/v24lyag-velocity-of-light-dr-edward-dowdye-extinction-shift-example-jason-verbelli.html

In that video, I discuss Dr. Dowdye's equations and the animated GIF from this page of his website dealing with the propagation and re-emission of light:

https://web.archive.org/web/20160331145545/http://www.extinctionshift.com/details03.htm

I made an interactive webpage that I'm still working on incorporating to a website at the time of writing this sentence. Here is a link to a video where I play around with the control sliders to change the velocity/frequency of the waves. There are three separate sine waves. The primary emission approaching the fixed window, the secondary emission propagating at the refractive index of the fixed window and then a third tertiary sine wave re-emitted/exiting the fixed window. The frequencies of all three sine waves are always in phase and automatically adjust. The velocity of sine wave #3 is always c. The velocity of sine wave #2 is always c/n. The velocity of sine wave #1 is variable. I took of the Re-Emission Visualization from Dowdye's animated GIF and refined it a bit more in the interactive page.

Re-emission Visualization - Extinction Shift Principle:

https://rumble.com/v4zl9i6-re-emission-visualization-extinction-shift-principle.html

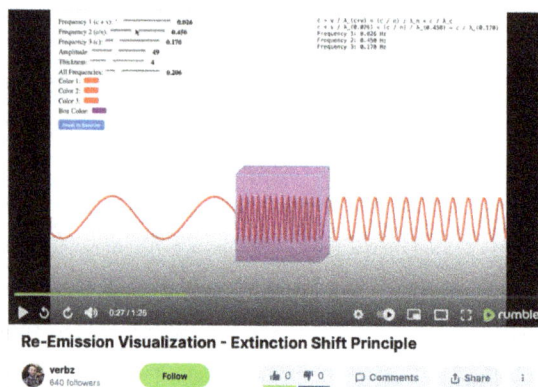

Fig. 4.6

# Humphrey Lloyd's Mirror Experiment... with a Twist

Humphrey Lloyd performed an optical experiment which produced interference patterns similar to Thomas Young's double slit experiment or Augustin Fresnel's experiments. But rather than using light through slits, Lloyd's setup uses a monochromatic light source and a mirror on the floor in front of the wall where the patterns emerge. This simulates the light AS IF it came from two slits because of the "virtual source" of light (the reflection from the mirror). This experiment provides further evidence for a wave-nature of light with no need for a particle wave duality.

A monochromatic light is blocked with a sheet or something but has a pinhole to allow the light through like a spotlight. That light grazes the mirror at a particular angle of incidence in a way that the light from the pinhole simultaneously hits the wall and the mirror. But this method produces the inverse interference fringes. Rather than brightest in the middle, it's dark. The fringes flip 180° out of phase because of the mirror, which produces constructive and destructive interference.

The double slit experiment with light has a particle experiment *ANALOG* using a collimated beam of electrons from a thermionic emission source. There is no analog using a beam of particles to produce an interference pattern for Lloyd's mirror that I've found so far. Not in English anyway. I'm just not too sure that many people think about rotating mirrors and particle beams producing secondary electron emissions and interference fringes. You're missin' out if you don't, I'll tell ya. (Just imagine if someone like Elon Musk were to hedge his bets and fund some technologies and experiments to challenge Lorentz invariant models.) There are dozens of cool ideas.

One of them is adding a twist to Lloyds mirror experiment by combining the previous chapter of a spinning mirror with Lloyd's experiment. But it's obviously not technologically feasible to make a 26-mile diameter mirror that spins at relativistic speeds to produce redshifted interference patterns. So we'll file this gedankenexperiment away for now. Fun to think about though. Here are some attempted proofs though:

Tangential velocity:

$$v_t = \omega r$$

The effective velocity describing the rotation of the mirror:

$$v_{eff} = v + u + \omega r$$

Applying Galilean variance & Dr. Dowdye's ESP to re-emissions of redshifted light, $Z$:

$$Z = \frac{\Delta v}{v} = \frac{u + \omega r}{c}$$

Describing the wavelength of the re-emitted and redshifted light:

$$\lambda' = \lambda(1 + z) = \lambda\left(1 + \frac{u + \omega r}{c}\right)$$

Now accounting for interference patterns and fringes. First with phase interference:

$$\Delta \Phi = \frac{2\pi}{\lambda}(d_2 - d_1) = \frac{2\pi}{\lambda\left(1 + \frac{u + \omega r}{c}\right)}(d_2 - d_1)$$

And finally the Constructive and Destructive interference patterns and fringes:

Constructive Interference happens when the phase difference is an integer multiple of $2\pi$:

$$\Delta \Phi = 2\pi m \quad \text{for} \quad m \in \mathbb{Z}$$

Deconstructive interference happens when the phase difference is an odd multiple of $\pi$:

$$\Delta \Phi = (2m + 1)\pi \quad \text{for} \quad m \in \mathbb{Z}$$

Interesting to note that Herbert Ives, Hendrik Lorentz and Albert Michelson all greatly opposed relativity until they day they each of them died. They refused to acknowledge relativity. If only they had known they were all correct but misinterpreted the interferometer results.

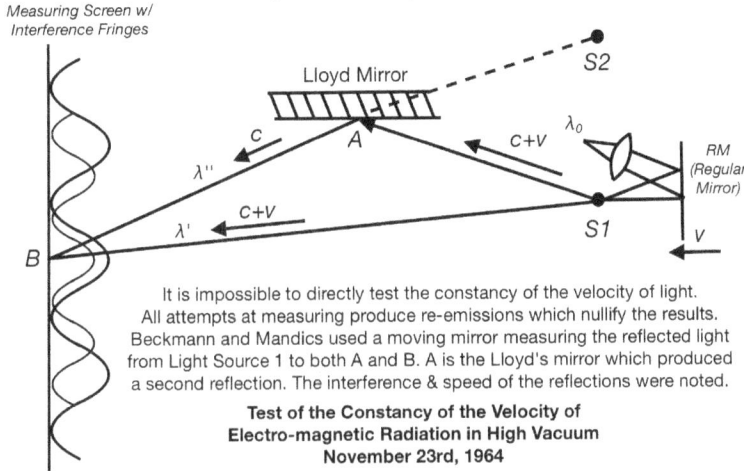

**Nullified Results of**
**Petr Beckmann &  Peter Mandics**
**Moving Mirror Experiment**

Measuring Screen w/
Interference Fringes

Lloyd Mirror

S2

$\lambda_0$

RM
*(Regular Mirror)*

$c$

$A$

$c+v$

$\lambda''$

$c+v$

$\lambda'$

S1

$V$

$B$

It is impossible to directly test the constancy of the velocity of light.
All attempts at measuring produce re-emissions which nullify the results.
Beckmann and Mandics used a moving mirror measuring the reflected light
from Light Source 1 to both A and B. A is the Lloyd's mirror which produced
a second reflection. The interference & speed of the reflections were noted.

**Test of the Constancy of the Velocity of**
**Electro-magnetic Radiation in High Vacuum**
**November 23rd, 1964**

Fig. 4.7

Using Dr. Dowdye's Extinction Shift Principle from Figure 8 of his book on the Lloyd Mirror, we can see that Galilean variance accounts for the same nullified results and interference fringes. Both reflections of light sources S1 and S2 interfere to produce the fringe patterns. The light from Source 1 to both A and B propagates at the velocity of $c + v$. The

wavelength of reflected light from A to B from S1 is: $\lambda' = \lambda_0 \left(1 + \dfrac{v}{c}\right)^{-1}$

$\lambda_0$ is the uninterfered-with wavelength of the re-emitted wave reflected by the RM. The wavepacket propagating along S1 to A gives rise to the tertiary reflection of A to B.
The velocity of the light packet from A to B is propagating at the velocity of c relative to the Lloyd mirror. The wavelength of the packet from A to B is:

$$\lambda'' = \lambda_0 \left(1 + \frac{v}{c}\right)^{-2} \text{ since the reflection of } \lambda_0 \left(1 + \frac{v}{c}\right)^{-1} = \left(1 + \frac{v}{c}\right)^{-2}$$

Meaning, there can be no Doppler shift or change of fringe shift pattern even if a source is moving toward or away from the interferometer. A primary is extinguished and re-emitted.

From page 24A of Dowdye's Extinction Shift Principle book, we can see the phase of the interfering light differs between Source 1's light straight to B versus Source 1's light reflecting from the Lloyd Mirror, A, then to B. The difference is:

$$\Phi = \Phi(\lambda_0, v) \quad \text{and} \quad \Delta\Phi = \frac{d}{dv}\Phi(\lambda_0, v)$$

The combined interference is:

$$\Phi(\lambda_0, v) = \Phi_{S1B}\left[\lambda_0\left(1+\frac{v}{c}\right)^{-1}, c+v\right] - \Phi_{AB}\left[\lambda_0\left(1+\frac{v}{c}\right)^{-2}, c\right] - \Phi_0$$

$\Phi_0$ is constant. The re-emission from each point of reflection yields a new wavefunction with the exponent of negative 2. The different frame of reference extinguishes the primary or preceding velocity. The point of the mirror A or RM from the Source 1 produces a brand-new packet of light from the frame of reference of that point. So there's no way to measure the original velocity at the screen, B since it's never the same light making it there. It's re-emitted from each frame of reference along the way, nullifying any intended results. Just because you don't observe a change of fringe patterns doesn't mean the velocity didn't change. Since the re-emissions are in phase with the incident primary, the fringes remain unchained. But that shift in frequencies IS a literal change in the velocity of light. That's the issue. People are interpreting the results under the assumption of an invariant Lorentz or Galilean framework for the velocity of light. The results are misinterpreted under false assumptions Below is the Beckman-Mandic interferometer with spinning mirror:

FIGURE 2. View of the interferometer assembly.

Fig. 4.8

It is never the same light that distorts or bends. The same light does not redshift or Doppler shift and then continue on within the same frame o f reference as a distorted primary. Light does not stretch like Rick James on Eddy Murphy's couch.

Fig. 4.9

Woow, that's quite the stretch Einstein...  I didn't know you were into Yoga!

Fig. 4.10

# Testing the "Rightness" of Relativity - Laser & Mirror

Point a laser or flashlight at a mirror.

Pointing a light at a mirror, does not give you any math or proofs. But when you point a laser at a mirror you can interpret that scenario under Lorentz invariant transformations. That route will mathematically "prove" it is the same light ricocheting off the surface of the mirror and continuing on within the same frame of reference.

However, that same scenario under Galilean variant transformations will mathematically "prove" it is a primary light hitting the mirror. The electrons making up the surface of the mirror will absorb that incoming primary and re-emit a brand new secondary light as an equal and opposite reaction. And that happens at the rate of c, serving the illusion that it's the same light bouncing back. Every point of interference and re-emission is a new frame of reference.

You can point a flashlight at a mirror 10,000 times... that experiment will not prove either scenario. Pointing a flashlight at a mirror does not prove it is the same light or a different light. That "proof" is in the mathematical interpretation of the scenario. Proof is math.

The entire world can conduct 10,000 experiments the same over and over... All it will do is provide evidence. And the entire world is interpreting that evidence under one specific mathematical framework of Lorentz invariance and relativistic procedures.

But indeed there are other experiments to show unreconcilable discrepancies of relativity; as well as, other mathematical approaches to yield the same predictions and solutions as relativity using Galilean variance.

Here is a video of a test with a laser and a mirror to show the impossibility of an experiment "proving" a theory.

https://x.com/TheRealVerbz/status/1788284452843692208

# — CHAPTER 5 —

# Johann Goethe - Light, Dark and Color

Most people are quite familiar with Sir Isaac Newton's rainbow color distribution from shining white light through a prism. Most people know it from the Dark Side of the Moon Album Cover by the band Pink Floyd.

Fig. 5.1

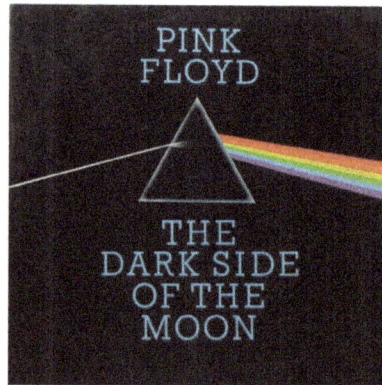

Fig. 5.2

But that is an erroneous depiction in textbooks of how a prism is supposed to "split" the white light both within the prism and directly exiting the prism. It's depicted that the colors of ROYGBIV (Red, Orange, Yellow, Green, Blue, Indigo, Violet) evenly distribute and are visually represented like a rainbow laser beam as soon as the white light is "filtered" or separated by the glass of the prism. And they treat all of that light as "the same light."

Johann Wolfgang von Goethe (1749-1832) was a German scientist who refined Newton's model of prisms and color about 100 years after Newton was around. Goethe found that color is a phenomenon that emerges at the boundary of light and darkness. That, for every primary color, there is a shadow color or a counterpart. Goethe (pronounced Ger-tuh) said that the ROYGBIV light distribution from the Newtonian spectrum emerges only after a certain distance away from the prism. Not directly out of the prism or within the prism as depicted in textbooks or Pink Floyd's Dark Side album cover. (It's actually ROYGCBV) Red, Orange, Yellow, Green, Cyan, Blue, Violet.

Newton played around with a beam of light within a darkened room. He blocked the direct sunlight with darkened curtains and cut a hole in the curtain to allow a conical ray of light to shine through. The light projected a circular spot of illumination on the wall. Isaac Newton held up a prism in the beam of light and the glass prism casted colors on the wall.

(Disco Time)

Fig. 5.3

But the ROYGCBV colors don't appear on the wall unless you hold the prism a certain distance away. Because it's only when the Red-Yellow and Blue-Violet merge after a certain distance do the other primary colors emerge. Goethe performed experiments in both light and darkness. Goethe's approach compliments, refines and completes Newton's approach.

One thing I've never seen and have wanted to do for years is to heat up a glass prism until it's red hot. Then perform the same experiment with the color on the wall to see how the refractive index of the glass changes as it cools. And how the colors change as the glass cools down to room temperature.

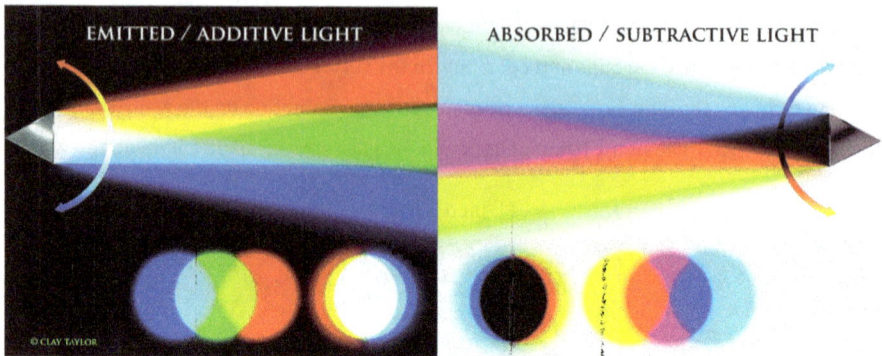

EMITTED / ADDITIVE LIGHT          ABSORBED / SUBTRACTIVE LIGHT

© CLAY TAYLOR

Fig. 5.4

Below is a chart I put together for harmonic frequencies correlating with musical notes and specific wavelengths of color. It's easy to find harmonic numbers. The rules are simple.

Just start with the number 1. Every time you double the number, you go up one octave within that note. Every time you triple the number, you bump up to a new note.
Like: "*Do, Re, Mi, Fa, So, La, Ti*"
When looking at the chart, starting from the upper left and moving in the column down, you double the number.

1

2

4

8

**etc.**

Double down for octaves of that note. Then you triple the number across the rows for a new note on the scale.

**1, 3, 9, 27, etc.**

Double Down, triple across. (I first heard this notion from personal talks with Joseph Cerecedes about his book, *Biological Resonance: Thriving in a Radioactive Universe.*)

Now keep doubling up the octaves anywhere from 39 to 49 times until you reach the terahertz range of visible light. Then find the wavelength associated with that frequency.

$$c = \lambda f$$

Find the wavelength $\lambda$ by using $\lambda = c/f$

Divide the long numbers in hertz to get the nanometer wavelength of color.

Look up the HTML Hex Code best associated with that wavelength of color.

Also take note there is a colored shadow counterpart for each listed wavelength.

1 doubled 49 times = 562.949953421312 terahertz:
Green Light – 533nm - Octave of C Note
Hex code is approximately #00ff00" to "#69ff00

3 doubled 47 times = 422.212465065984 terahertz
Deep Red Light/ Near Infrared Light – 710nm - Octave of G Note
Hex code is approximately #8b0000 to #ed0000

9 doubled 46 times = 633.318697598976 terahertz
Blue Light – 474nm - Octave of D Note
Hex code is approximately #0000ff or #00bbff

27 doubled 43 times = 474.989023199232 terahertz
Deep Red Light – 631.5nm - Octave of A Note
Hex code is approximately #ff4600 or #ff0000 or #e50000

81 doubled 43 times = 712.483534798848 terahertz
Violet Light – 421nm Octave of E Note
Hex code is approximately #8b00ff

243 doubled 41 times = 534.362651099136 terahertz
Green Light – 561.7nm Octave of B Note
Hex code is approximately #c9ff00

729 doubled 39 times = 400.771988324352 terahertz
Red Light – 400.7nm Octave of F Note
Hex code is approximately #FF0000

729 doubled 40 times = 801.534976648704 terahertz
Near-Ultraviolet Light – 801.5nm Octave of F Note
Hex code is approximately #8A2BE2

| C | G | D | A | E | B | F |
|---|---|---|---|---|---|---|
| 1 | 3 | 9 | 27 | 81 | 243 | 729 |
| 2 | 6 | 18 | 54 | 162 | 486 | 1458 |
| 4 | 12 | 36 | 108 | 324 | 972 | 2916 |
| 8 | 24 | 72 | 216 | 648 | 1944 | 5832 |
| 16 | 48 | 144 | 432 | 1296 | 3888 | 11664 |
| 32 | 96 | 288 | 864 | 2592 | 7776 | 23328 |
| 64 | 192 | 576 | 1728 | 5184 | 15552 | 46656 |
| 128 | 384 | 1152 | 3456 | 10368 | 31104 | 93312 |
| 256 | 768 | 2304 | 6912 | 20736 | 62208 | 186624 |

| Color | Wavelength Interval | Frequency Interval |
|---|---|---|
| violet | ~ 430 to 380 nm | ~ 700 to 790 THz |
| blue | ~ 500 to 430 nm | ~ 600 to 700 THz |
| cyan | ~ 520 to 500 nm | ~ 577 to 600 THz |
| green | ~ 565 to 520 nm | ~ 526 to 577 THz |
| yellow | ~ 590 to 565 nm | ~ 508 to 526 THz |
| orange | ~ 625 to 590 nm | ~ 484 to 508 THz |
| red | ~ 740 to 625 nm | ~ 405 to 484 THz |

Fig. 5.5

Interesting to note that only numbers 2 and 3 in this chart are prime numbers. The harmonic Newtonian spectrum doesn't seem to have primes that I've found so far.

| C | G | D | A | E | B | F |
|---|---|---|---|---|---|---|
| 1.4 | 4.35 | 6.99 | 68.9 | 59.84 | 324.82 | 1113.7 |
| 2.8 | 8.7 | 14 | 137.8 | 119.68 | 649.64 | 2227.3 |
| 5.6 | 17.4 | 28 | 275.6 | 239.36 | 1299.28 | 4454.6 |
| 11 | 34.8 | 55.9 | 551.2 | 478.72 | 2598.56 | 8909.3 |
| 22 | 69.6 | 112 | 1102.4 | 957.44 | 5197.12 | 17819 |
| 45 | 139 | 224 | 2204.8 | 1914.88 | 10394.2 | 35637 |
| 90 | 278 | 447 | 4409.6 | 3829.76 | 20788.5 | 71274 |
| 179 | 557 | 895 | 8819.2 | 7659.52 | 41577 | 142548 |
| 358 | 1114 | 1789 | 17638.4 | 15319.04 | 83153.9 | 285097 |

Fig. 5.6

But here in the Goethe Spectrum we can see there are 7 primes already.

11, 139, 179, 557, 1789, 17819, 285097

Negative Color Wavelengths HTML Hex Codes:

C note # 8D0BB5

G note # 24CFE0

D note # AD4705

A note # 2DD3E1

E note # 80D50A

B note # 2A03AE

F note # 15CCDC and 81C524

Green exists in the prism experiments when the red-yellow and cyan-blue merge to produce the green. There is also magenta in the prism experiments when you use 2 prisms. Green does not exist immediately out of the prism. Only when the other colors merge through distance and dispersion.

Magenta is not a real WAVELENGTH, but it is a real color.

Magenta is the shadow of green. Magenta doesn't exist in the electromagnetic spectrum unto itself like the Newtonian spectrum. There is a color code equivalent but not a wavelength. It's like your brain fills in the negative and produces a remnant shadow at a constant which you interpret as magenta.

What I did in the example on the following pages was use a prism to cast the "rainbow" on the floor. I put a piece of white poster board on the floor where the prism was casting the re-emitted colors. I then placed a second prism standing vertically where the colors were cast on the poster board. I shifted the position of the prism, spun it on its axis and found beautiful colors. I took the images on the following pages on 06/07/2024 at 5:30pm PST on an iPhone 8 camera in San Diego, CA. I uploaded a video of the shifting colors in real time as well. For those on kindle of reading on the net, click this link:

For those reading a hard copy of the book... just keep tapping your finger on the paper over and over... it'll work eventually... I swear. 😉

So, the Newtonian frequency and wavelength of color associated with a harmonic of sound bumped up to the terahertz equivalent can be expressed in actual wavelengths. But the Goethe spectrum must be expressed in hex code equivalents because there is no wavelength for some of the colors in the Goethe spectrum. It's the shadow OF the Newtonian spectrum. Not wavelengths unto themselves.

Dark Side of the Moon? How about, Light Side of the Sun.

Roger Waters, eat your heart out!

Fig. 5.7

Fig. 5.8

Fig. 5.9

Fig. 5.10

Fig. 5.11

Fig. 5.12

Another experiment is to have mirrors deflecting a ray of sunlight onto a prism, which is already distributing light. Have direct sunlight entering your window, projecting onto the floor or the wall. Cover up the window with cardboard and block the sunlight. Poke a hole in the cardboard and allow the sunlight to come through the pinhole. Let's say the light project from the pinhole onto the floor. Place a prism in the spot of sunlight on the floor. Somewhere in the room, you will see the red-yellow and cyan-blue with white in between. And depending how far away, the prison is two, the wall or surface, where the color is projected... that distance will determine if the red yellow and cyan blue merge to produce the rest of the rainbow colors.

Now take a mirror and shine the light on to the prism, where it is already sitting on the spot from the sunlight through the pinhole. Project another ray of light from another window on the prism to see how the colors react. Slowly spin the prism or tilt it and see how the colors distribute.

Then place more prisms in the beams that are shined onto the floor. And see what colors those produce. Get a magnifying glass, magnify, a specific, isolated color to a focal point, and see what happens through a prism as well. Or magnify the isolated color from a prism and spin the magnifying glass. I spent the different prisms at different rates and move them around at the same time or in combinations.

There are many different experiments to try that haven't been done before.
Pehr Salstrom is a great resource for experiments and learning how Johann Goethe's model compliments Isaac Newton's model. Two other invaluable resources on this topic are from my good friends Clay Taylor and Thomas Joseph Brown. Another great reference is a huge compilation of color theories in a two-book set called  *THE BOOK OF COLOUR CONCEPTS* by, Alexandra Loske and Sarah Lowengard.

# Gravitational Shadows

I think there is a similarity between Illumination and Gravity. It doesn't matter if a room is lit by a dim red light or a bright blue light or white light, etc. The effect of *any* range of illumination allows you to observe the room in real time. Just as it doesn't matter if a body is emitting different spectra of "gravitons"/ emissions. The effect of *any* range of gravity allows other bodies to be influenced. Looking at gravity as being binary or "on or off" or positive/ negative doesn't cut it. But imagine if someone wasn't able to see ANY color. Just that illumination allows the observer to see in real time. This is how the world views gravity currently. No discernment between wavelengths/ frequencies OF gravity. Gravity follows the same rules as light. And the world still has misconceptions about light, still using an outdated invariant model under Lorentz.

The double down, triple across rule does not apply to the Goethe spectrum and hertz frequencies. It's a totally different dynamic because the colors of the Goethe spectrum are not wavelengths unto themselves.

How can color not have a wavelength? The colors in the Johann Goethe Spectrum aren't exactly wavelengths of light unto themselves. They are the SHADOWS of the Newtonian Spectrum.

I have a theory on that regarding negative space or the brain filling in what is not there. And the brain interprets the gaps as color sometimes. When white light hits the surface of a red ladybug, the white light is absorbed by the electrons making up the red elytra (wing coverings). The electrons making up the elytra re-emit every color EXCEPT red into the eye. The brain fills in what is not there, and we interpret that as color for that scenario. We interpret the difference of what is there versus what is not.

**Every color BUT Red
Re-Emits into the Eye.**

white light coming in

red light

red surface

white light coming in

red surface

**Your brain interprets the Negative many times.
Fills in the gaps of what is NOT there. You "see" the
difference between what IS there and what is Not.
(unless you are looking at a monochromatic or
coherent light. Then you are seeing the Only thing
that Is there in relation to everything that is Not.)**

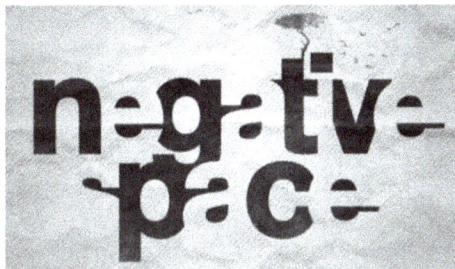

negative space

Fig. 5.13

And contrast to that... if we observe a coherent light from a laser or monochromatic light from an LED... the brain doesn't need to fill in any gaps because the eye sees the only thing that is there in relation to everything that is not. Therefore, a red coherent light will be interpreted by the brain, not as everything but red... but Only red. The brain is focusing on what is missing, otherwise.

But when you are seeing the red from your shirt or the surface of an object... that is because the white light hitting the object is being absorbed by the electrons making up that object and then a brand-new light is re-emitted into your eye. Your brain then interprets what is missing as the red color in that scenario. Tricky, tricky!

Fig. 5.14

Fig. 5.15

Same premise with the blue hue from seeing the surface of objects like a morpho butterfly wing verses a blue laser beam or monochromatic LED. When you see the blue wing, you're interpreting the blue that isn't there. When you see the blue laser, you're interpreting the only color that is there, blue.

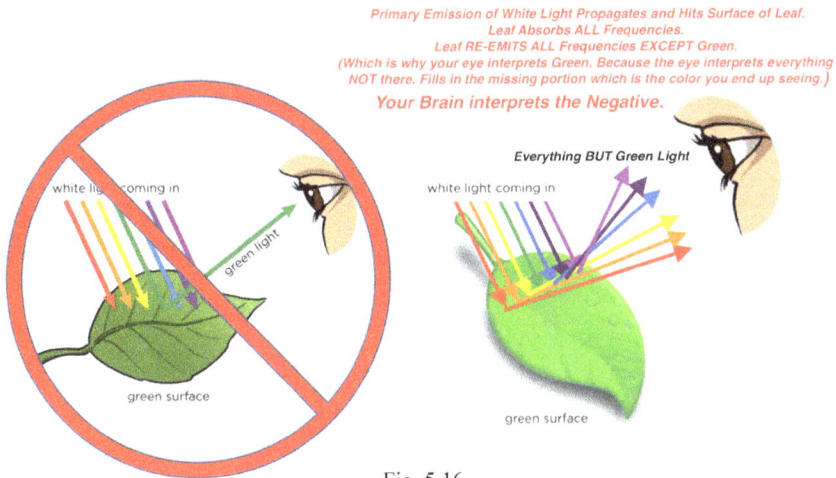

Fig. 5.16

This means the Newtonian frequency and wavelength of color (RGB) associated with a harmonic of sound bumped up to the terahertz equivalent can be expressed in actual wavelengths. But the Goethe spectrum (CMY) must be expressed in hex code equivalents because there is no wavelength for the Goethe spectrum per say. It's the shadow OF the Newtonian spectrum. Not wavelengths unto themselves.

But I tried anyway to take the negative of the Newtonian spectrum the chart and then find the html color hex code that correlates with that and apply that to frequencies in hertz. But it just doesn't work like that mathematically or in reality in regard to color.

You can find an equivalent color code but not a literal wavelength in nanometers or a frequency associated with that wavelength. Because colors like magenta do not exist as a literal wavelength. Again, magenta is the shadow of green.

It took me a while to realize all that, but I made a chart after discovering the negative of the Newtonian chart... getting the hex codes... finding the wavelength best matching that... convert the wavelength to frequencies... then divide by 2 over and over until getting back down to the audio hertz range again. Just the inverse of how I got the Newtonian colors for the chart. But the hertz frequencies in this chart, again, do *not* match up with the Goethe spectrum because there is no wavelength or frequency for magenta and certain Goethe colors.

Therefore... assuming gravity behaves like light... we can make a logical leap that gravitational shadows are most likely a thing and are real. And as a consequence of the model, there must be certain "shadows" to gravitational wavelengths and frequencies that could not be found as a literal gravitational influence but rather might be the shadow of a primary influence. A constant remnant shadow.

Just as light illuminates reality in the visible light range, regardless of color... gravity affects matter regardless of its wavelength. It doesn't matter if the scene is illuminated by a monochromatic star producing a red light, green light, or blue light. The light illuminates reality as it is in real-time. The consequences of this model show that light and gravity behave the same. All the light you ever see is ultimately the re-emitted light from the electrons making up yourself... and I think the same applies to gravity.

All the gravity we ever feel is ultimately the re-emitted gravity from the electrons making up ourselves. And the illusion is that we can only measure the re-emitted gravity from our own electrons, therefore it is physically impossible to measure the primary gravity from any source. It is impossible to measure any plane wavefront gravitons because the gravity is re-emitted as secondary spherical wavefronts from the frame of reference of the electrons making up Earth, and the electrons making up the electrons comprising everything on the Earth. The plane wavefronts of gravitation are coherent and the spherical wavefronts are incoherent or monochromatic. These gravitational waves have nothing to do with the fictional gravitational waves claimed by LIGO and ripples in space-time.

So, it's my personal belief, that according to the consequences of this theory under Galilean variance and Dr. Dowdye's *Extinction Shift Principle* that light and gravity behave the same. By that logic, there are probably gravitational shadows. I think gravity has a spectrum just like light. And just as different colored light has a different colored shadow counterpart... I think gravity and re-emissions of gravity and gravitational shadows will affect surrounding bodies. Like the bodies produce secondary echoes or gravitational remnants.

Gravity is thought of as Newtonian in classical physics. So, if there's a Newtonian spectrum of light and color... and a Goethe spectrum of light and color... and gravity behaves like light... then perhaps the shadow of a Newtonian gravitational emission could be called "Goethenian Gravity." And by the logic and consequences of Galilean variant transformations, when Newtonian and Goethenian gravity merge, it might produce a full spectrum of gravity. Together brother.

Fig. 5.17

# Indirect Gravity & Gravitational Shade

There is direct sunlight and indirect sunlight. Indirect light is being in the shade or sitting under a tree in the tree's shadow. The tree is eclipsing the light. The sun produces white light... the opposite of white is black; therefore, all objects that aren't translucent cast a black shadow. Block/eclipse sunlight with stained glass and you get the colors of the light projecting around the room in beautiful arrays. The consequences of Galilean variance show that gravity behaves like light. If light creates a shadow, then so must gravity by this logic. So, we will work under this assumption based on the consequences and justification of Galilean variant transformations. If you consider light as being magick... then consider this chapter as "shade magick."

Fig. 5.18

The shadow is cast as a counterpart 180° to the direction of the light source.
And this holds true as we see the darkest part of a shadow is 180° from the source. The darkness of the shadow drops off to grey like concentric layers of an onion or a Russian doll away from the darkest point.

Likewise, this holds true for color as well. On a color wheel, we can see the shadow of red is cyan which are situated 180° apart. The shadow of green is magenta, and indeed they are 180° apart. The same for blue and its counterpart shadow yellow.

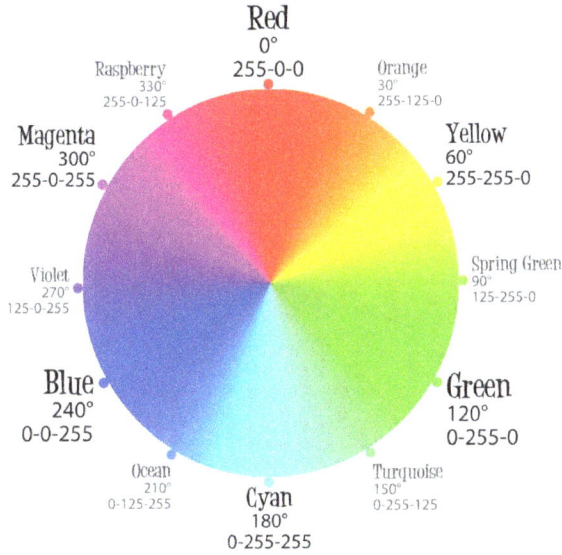

Fig. 5.19

In a room, you don't know there is a shadow being cast unless there is a floor or wall upon which to cast it. But indeed, if you block light, it will cast a shadow, even in the void of space. In the void of space though, there are no walls or floors upon which to cast the shadow. A planet eclipsing a distal light source will cast a shadow upon the nearest celestial bodies acting as a backdrop for the shadow to be cast.

Some flowers and plants prefer being in the shade while others prefer direct sunlight. But is it a black shadow the plants in indirect sunlight are really after? Or would some plants prefer specific isolated Goethe colors and combinations? Perhaps it's not just the heat that is the reason for some plants not liking to be in direct sunlight but rather the conflicting frequencies of the white light, so it needs the counterpart shadow or aspects of it.

If you have a monochromatic light, it will cast a monochromatic counterpart shadow. Block blue primary light with a sphere and it will result in a yellow shadow on the ground opposing the eclipsed physical solid body and the primary emission cast onto it.

Einstein really needed to do more internal reflection and shadow work.

Fig. 5.20

Wanting help from graphic designers so I can animate these depictions, I set up a crude central planet and three equal distant lights. We can see the shifting of those colors as an observer from that central planet while turning on/off the lights in different combinations.

Fig. 5.21

On the following page, we can see the different colored shadows the lights produce.

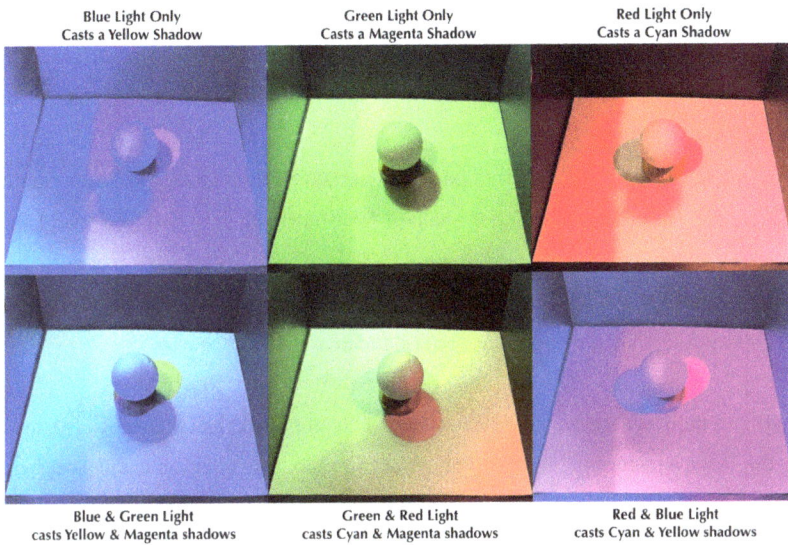

Blue Light Only
Casts a Yellow Shadow

Green Light Only
Casts a Magenta Shadow

Red Light Only
Casts a Cyan Shadow

Blue & Green Light
casts Yellow & Magenta shadows

Green & Red Light
casts Cyan & Magenta shadows

Red & Blue Light
casts Cyan & Yellow shadows

Fig. 5.22

Blue light produces a yellow shadow. Green light produces a magenta shadow.

And red light produces a cyan shadow. Then combinations of all 3 simultaneously.

We can see the cyan, yellow and magenta... and when those intersect... we get the original Newtonian primary colors again. But in order to observe all the colors and their combinations together... it requires yet *another* white light to illuminate *that* whole scene.

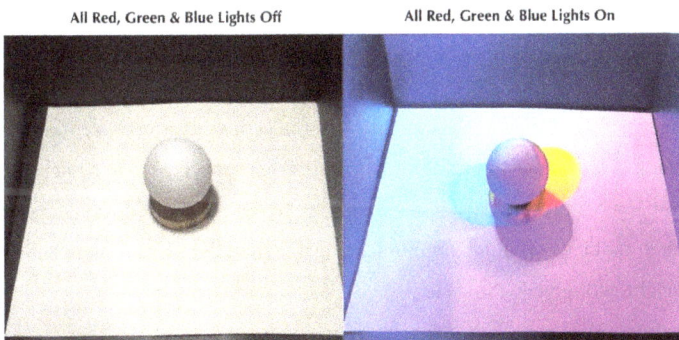

All Red, Green & Blue Lights Off

All Red, Green & Blue Lights On

Fig. 5.23

So, if a planet is eclipsing the incoming primary gravity and light, you will still be subject to its counterpart shadow on the opposing side. A shadow can travel across a surface as quickly as the electrons can re-emit light. The "redshift" of shadows can take place as well.

Notice that a shadow cast onto a wall will seem to "travel" much faster the closer your hand is to the source. Shine a flashlight on the ceiling. Now block the light with your hand about a foot away from the light and shake your hand back and forth. The shadow cast onto the celling by your fingers will wave. And the speed at which the shadow travels across the ceiling will be determined by how close your hand is to the light source.

It can get to a point where the shadow cast onto a distant object can theoretically travel across the surface faster than light or a faster than a radio transmission could sweep that same distance. The shadow cast onto a distant source can traverse the path of a body faster than the wavefront of a primary emission can traverse. If you consider direct light a 1 and a shadow a 0, then data could be transmitted faster than light via a projected shadow's rate of change sweeping across a given area. You could at least get someone's attention faster with a flickering shadow, eclipsing existing light before a radio transmission reached the same distant location.

The point is that the rate at which a gravitational shadow can be produced across the surface of a distant body from a source being eclipsed is analogous to light which abide by specific rules of optics. The slight action of your fingers over the bulb of the flashlight can produce a shadow traveling many times faster across the ceiling.

An eclipsing is different than a lack of light since the eclipse produces a counterpart shadow. A lack of light does not have a shadow. It's just the absence of a packet altogether. But a shadow is a counterpart of a given light 180° to the eclipsed source of a given wavelength. In theory, we could communicate faster than light by lifting our pinky finger in front of a bright enough light source.

Over a hundred years of thought experiments attempting to communicate faster than light, thinking it would violate causality, and here the tip of our pinky finger might be able to demonstrate more than what 100 years of relativistic experts said was impossible.

This reminds me of a quote from Paracelsus.
"Let me tell you this: every little hair on my neck knows more than you and all your scribes, and my shoe-buckles are more learned than your Galen and Avicenna, and my beard has more experience than all your high colleges." -- Paracelsus, also known as Philippus Aureolus Theophrastus Bombastus von Hohenheim. (C'mon, pick a name buddy!)

On the following pages are the basic premise and math to substantiate the dynamics of shadow propagation. The speed at which a shadow traverses a distant surface depends on the distance between the eclipsing body and the surface the shadow is cast upon.

I propose there is a gravitational analog. The effects would be analogous to an optical shadow, but the attraction/repulsion factor of bosons to fermions and primary to secondary gravities would remain unaffected since the source itself remains unchanged. It's just the same signal being eclipsed which still produces a frequency shift in phase to the primary stimulus. The same effects of gravitation apply at a constant regardless of primaries or secondaries, direct or indirect gravity.

This would imply that the effects of gravitational re-emissions would have immediate effects within an already existing gravitational field. The frequencies would be in phase across great distances. A shadow's rate of change can traverse the surface of a body faster than an initial wavefront of a given emission propagates in the same amount of time. But the shadow would be cast within an already existing gravitational field just as a shadow is cast within an already existing state of illumination. The constant of the emission would remain unaffected to the bodies within that primary gravitation or the counterpart shadow.

In this equation, $d$ is the distance from the primary source to the eclipsing body. D is the distance from that eclipsing object to the surface upon which that shadow is cast. And the speed at which the shadow travels across the surface is:

$$V_s = v \cdot \frac{D}{d}$$

(V by itself is the speed of the eclipsing object. $V_s$ is the velocity of the shadow)

The farther the distance, D relative to d, the more the $V_s$ can be much greater than the velocity of light, C. The gravitational analog might look something like:

$$V_{gs} = \frac{c \cdot D}{d}$$

(The velocity of the gravitational shadow is equal to the velocity of light constant times the distance of the body being cast upon, divided by the distance the eclipsing body is from the source of gravity.)

Whether a flower is in the shade or not... it's still illuminated at a constant.
Whether a body is in the gravitational shade or not... it's still affected by its host star at a constant.

Take note, the shadow of a coherent light produced by either a laser, cooper pairs decoupling and emitting photons or coherent gravitons... that gravitational shade or gravitational shadow is different than the re-emission of an incoherent/polychromatic light or incoherent gravity. And all of that is also different from a monochromatic light or theoretical "monochromatic gravity." "Monochromatic" obviously being analogous to the fermion emission of photons counterpart to the boson emission of gravitons within a certain range. The packets would be emitted and re-emitted in a phase-coherent manner. Perhaps maintaining their synchronized wavefronts relative to source and absorber from medium to medium, as Walter Russell implied.

Math is a tool... and a language. So, pardon my French:

# Coherent Sources

**Wave Function for Coherent Light:**

$$\Psi_{\text{coherent}}(\mathbf{r}, t) = A e^{i(\mathbf{k}\cdot\mathbf{r}-\omega t)}$$

Interference pattern:

$$I(\mathbf{r}) = |\Psi_{\text{coherent}}(\mathbf{r}, t)|^2$$

**Wave Function for Coherent Gravity:**

$$\Phi_{\text{coherent}}(\mathbf{r}, t) = B e^{i(\mathbf{q}\cdot\mathbf{r}-vt)}$$

Interference pattern:

$$J(\mathbf{r}) = |\Phi_{\text{coherent}}(\mathbf{r}, t)|^2$$

# Incoherent Sources

**Wave Function for Incoherent Light:**

$$\Psi_{\text{incoherent}}(\mathbf{r}, t) = \sum_n A_n e^{i(\mathbf{k}_n\cdot\mathbf{r}-\omega_n t)}$$

Intensity Pattern:

$$I_{\text{incoherent}}(r) = \sum_n |A_n|^2$$

**Wave Function for Incoherent Gravity:**

$$\Phi_{\text{incoherent}}(r, t) = \sum_m B_m e^{i(q_m\cdot r-v_m t)}$$

Intensity Pattern:

$$J_{\text{incoherent}}(r) = \sum_m |B_m|^2$$

# Monochromatic Sources

**Wave Function for Monochromatic Light:**

$$\Psi_{\text{monochromatic}}(\mathbf{r}, t) = A e^{i(\mathbf{k}\cdot\mathbf{r}-\omega t)}$$

The shadow intensity:

$$I_{\text{monochromatic}}(\mathbf{r}) = |A|^2$$

**Wave Function for Monochromatic Gravity:**

$$\Phi_{\text{monochromatic}}(\mathbf{r}, t) = B e^{i(\mathbf{q}\cdot\mathbf{r}-v t)}$$

The shadow intensity:

$$J_{\text{monochromatic}}(\mathbf{r}) = |B|^2$$

Gravity obviously does not have a visible spectrum or a color, but we can make analogies to light. I pose that anything light can do... gravity can do. It's an emission that behaves just like light even though we can't see it. But rather than illumination, it's gravitation. As strange as it sounds, the laws of optics apply to gravity even though we can't see gravity.

The idea is to get a sense that gravity has a spectrum just like light, since both light and gravity are emitted by electrons. Different gravitational wavelengths and frequencies result from being emitted by decoupling cooper pairs/ bosons. But again... it is crucial to not get stuck in the quantum interpretation. We must drop and totally abandon all interpretations of relativity and QED. Neither photons nor gravitons are particles. But that "wave" people refer to is in regard to the absorption and re-emission by bosons and fermions in a wave-like manner or pulsations. Do not get misled by visualizations in university textbooks attempting to "oversimplify" this concept. They use depictions of literal rippling capillary waves at the surface of water or at the boundary of a condensed matter medium.

De Broglie waves at the surface of water resulting from an impact against the meniscus cannot be compared to absorption and re-emission at a constant rate (frequency). The difference in energy levels will cause interacting fermions to re-emit analogous to light in different counterparts like a shadow or remnant, or a reflection/ echo. All of these examples are re-emissions as a consequence of Galilean transformations for primary and secondary gravitons.

Shine a green laser beam at a cloud of mist, fog or smoke. The coherent laser light will illuminate the cloud. The particles in the cloud are acting like a projector screen or a backdrop for the light to shine onto. The beam of the laser itself is collimated, but the re-emitted light from the illuminated cloud still glows at the same wavelength as the laser. The light from that illuminated cloud can make a shadow if you block that light. The green light from the laser will show as a coherent magenta shadow when that light is eclipsed or blocked.

The formation of a complimentary coherent shadow can be expressed as the following:

$$I_{green}(\mathbf{r}) = I_0 \cos^2\left(\frac{\pi d \sin(\theta)}{\lambda}\right)$$

When you block the coherent green light, you get magenta. But magenta can be further broken down into its red and blue components as such:

$$I_{blocked\ green}(\mathbf{r}) = I_{magenta} = I_{red} + I_{blue}$$

I just arbitrarily listed the subscript below the $I$ variable as "blocked green". You could write $I_{bgreen}$ or $I_{blocked}$ or $I_{blg}$ etc. Same with the subscript coherence or $Variable_{coho}$ etc. You can put anything you want to denote the concept being accounted for.

$$Variable_{whatever\ you\ want\ to\ name\ it}$$

Now we will incorporate Dr. Dowdye's method while checking the equations are in thermodynamic balance as per Dr. Pierre-Marie Robitaille's principles. Dr. Robitaille discovered an axiom from Peter T. Landsberg's work in Thermodynamics with Quantum Statistical Illustrations, Interscience Publishers, New York, 1961: p.142. That work proposes a fundamental rule which can be known as **The Fourth Law of Thermodynamics**. This pertains to "*intensive and extensive properties.*" If an equation has a variable that is either extensive or intensive on the left, then the right side must be balanced in kind.

Following that rule, we can check to see if an idea and equation is mathematically sound as a proof. If the derivation inappropriately mixes and matches intensive and extensive properties, then the equation is invalid and should be retracted from scientific literature. (Or at least put those books in the science fiction section.)

Primary Coherent Graviton Emission:

$$\psi'_{\text{primary}}(\mathbf{r}, t) = A e^{i[(\mathbf{k}-\mathbf{v})\cdot\mathbf{r}-\omega t]}$$

Secondary Incoherent Graviton Re-Emission:

$$\psi'_{\text{secondary}}(\mathbf{r}, t) = \sum_n A_n\, e^{i[(\mathbf{k}_n-\mathbf{v})\cdot\mathbf{r}-\omega_n t]}$$

Primary Coherent Gravitational Field:

$$\Phi_{\text{primary}}(\mathbf{r}, t) = f\left(\frac{M}{r}\right) e^{i(\mathbf{k}\cdot\mathbf{r}-\omega t)}$$

Secondary Incoherent Gravitational Field:

$$\Phi_{\text{secondary}}(\mathbf{r}, t) = \sum_n f\left(\frac{M_n}{r}\right) e^{i(\mathbf{k}_n\cdot\mathbf{r}-\omega_n t)}$$

The Primary Coherent Gravitational Wavefront which is Planar:

$$\Phi_g(z,t) = A \sin(\omega t - kz)$$

The Secondary Incoherent Gravitational Wavefront which is Spherical:

$$\Phi_g(r, t) = \frac{A}{r} \sin(\omega t - kr)$$

The gravity imparted by the decoupling of cooper pairs/ bosons plays no role on the bosons themselves. Which is why I left out the gravitational constant of G. It would be inappropriate to place that in the equation because it would violate the 4th law of thermodynamics. The equation would not be in thermodynamic balance if adding the G. Gravity does not affect light or electromagnetic radiation whatsoever. That would be analogous to a system doing work on itself if gravity could affect the cause of itself. I will address that in a later chapter regarding claims of alleged gravitational lensing. One cannot introduce the concept of $M^2$ in equations either. It's erroneous and will be explained in a later chapter on Stephen Hawking's invalid equations.

But in regard to a gravitational shadow or a refractive analog to gravity might be shown as:

$$\delta\theta_s = \frac{4GM}{R_{eff}c^2}\,n$$

Where $n$ is the refractive index of the ionized medium around a body. If there is no plasma or medium present then there's no deflection of light, period.

If a superconductor or rotating superconductor has a gas bound to a convective flow within a boundary around the system, then the deflection of light would be dependent upon the refractive index of the medium. But that refractive index could theoretically change based upon the amount of coherent bosons present or a host of other factors and special circumstances. But in any circumstance, there *must* be some medium present to absorb and re-emit/deflect light since gravity has no effect on light. Gravity and light are both emitted by electrons; therefore, gravity and light have no effect on each other.

*That which is emitted by electrons cannot directly affect that which is emitted by electrons.*

# Deflection of Light

In 1915, Einstein wrote in a letter to the Times of London about three postulates under the Theory of Relativity. Albert Einstein seemed to insist that if any one of the three postulates proved to be wrong that Relativity would no longer be needed and would eventually collapse. One might even go so far as to argue that if *any* alternative methods were *ever* to be developed in the future that it would justify abandoning relativity for the simpler explanation. That time arrived in 1991, thanks to Dr. Dowdye.

The scientific community treats the corona and photosphere of the Sun as a vacuum. So, they attribute the deflection of light to gravity. But there is clearly condensed matter present in the form of metallic hydrogen in the photosphere and corona. Which means, the deflection of light at that location can be explained through refraction.

There is no deflection of light above 1 solar radius. The same equation for relativity bending light is the same equation for 3D optics under Fermat.

$$\delta\theta = \frac{4GM}{Rc^2}$$

But that equation, again, ONLY applies at 1 solar radius.
Relativity invents a bunch of other equations to suggest how light *SHOULD* bend a given distance away at 2 solar radii or 3 solar radii, but that has never been observed.

Gravity drops off according to the inverse square, but light does not deflect according to the inverse square. Luminosity drops off according to the inverse square, but light only deflects according to 3D optics under Fermat, via refraction. Following a least time path or minimum-energy path. **If there's no condensed matter or plasma present, then there's no deflection of light.** I'll reiterate that repeatedly.

The equation,

$$\Delta v = v_o \frac{GM}{Rc^2}$$

describes the change in velocity of re-emitted light due to the gravitational potential gradient. Light could be approaching a body that is producing gravity at $c + v$ or light could be receding away from a body at $c - v$. Whatever speed that light is relative to the body, we can label as $v_o$ which is the rest velocity.

The alleged gravitational lensing due to distant galaxies is different than the alleged gravitational lensing produced by the sun this page is this image address is not specifically here that is data processing errors from amplify light by over a factor of 1000 times brighter than it naturally is in space. That produces forward light scatter and optical distortions just in that specific wavelength due to flaws of the satellite itself.

In some situations, the amplification of light and data processing result in the streaks and arcs attributed to gravity bending light. However, according to GR, those optical distortions, streaks and arts should be seen equally throughout the entire spectrum, and in **all** wavelengths. But they are not. The distortions are frequency and wavelength dependent. Again, those artifacts only appear because they amplify light by over a factor of 1000 times. In some cases, it's debatable that the artifacts are purposefully manufactured by relativists in order to deviously acquire evidence to support their model.

The EHT team did something similar with the M87 ring of light as shown on the news April 10th, 2019. That image isn't real. I take it as fraud from data-processing the noise to produce artifacts which aren't there. They amplify the noise to produce an artifact that best matches their cartoon simulations of what they *want* to see. It's the epitome of confirmation bias. Scrutiny from Dr. Pierre-Marie Robitaille and others suggest the claimed image of M87 exceeds the capability of the radio telescopes by over a factor of 1250 times, so it's garbage in garbage out again (GIGO).

# Gravity Does Not Affect Light

As for the deflection of light (or lack thereof) by the Sun and Sagittarius A:

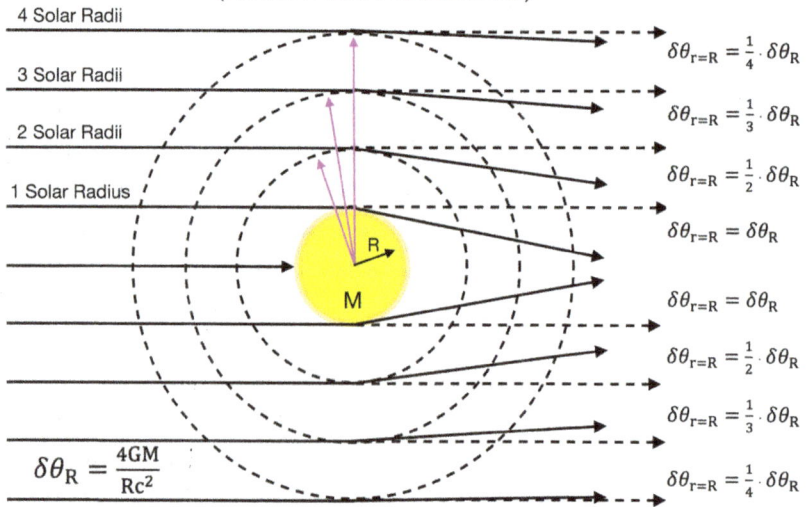

How the Sun's gravity is SUPPOSED to bend light at each distance away from its surface as depicted by Relativity (NEVER BEEN OBSERVED)

Fig. 5.24

According to Lorentz invariance, relativity does not discriminate between different wavelengths and frequencies when it comes to gravity bending light. It's an all or nothing phenomenon with relativity when it comes to the deflection of light. Yet the Sun only deflects microwaves at its surface at exactly 1.752 arc seconds. There is no deflection above 1 solar radius in reality, yet computer simulations and cartoons depict ALL light bending according to the inverse square away from the Sun's surface. (Yes, the Sun has a true surface, unlike what mainstream solar theory teaches.)

The same artifacts, bending of light, changing position of stars, "gravitational lensing" and distortions should all be seen equally in the radio spectra as they are in microwave or visible light, etc. The same bending should be equal throughout the entire electromagnetic spectrum according to relativity. Lorentz invariance does not get to pick and choose which wavelengths bend while others do not. So let's get specific about microwaves deflecting at the Gaussian surface of the Sun. Must mention once more that microwaves deflect ONLY at the surface of the Sun, at 1 solar radius. And that deflection is at exactly 1.752 arc seconds.

Dr. Edward Dowdye wrote an article on this topic specifically dealing with microwaves deflecting ONLY at the surface of the sun at 1.752 arcseconds. He submitted it for peer review. It was rejected by an anonymous referee. The peer review process is supposed to be public and documented by "peers". But this anonymous coward rejected the review while mischaracterizing the argument itself. Either gross negligence or purposeful subversion.

I, the Author, submitted to the Journal *Astronomical Notes* contact information on Four (4) renown experts on the subject matter of the published paper. I submitted E-Mail addresses and phone numbers to the Journal at their request. The Journal shamefully did not contact any of these experts.

You may read this Referee's report for yourselves. The Journal didn't even have the courtesy to give me the name of this chosen Referee.

The Referee chooses to use Microlensing and the subject matter dealing with the so-called *OGLE team*, *OGLE* designated to mean *"Optical Gravitational Lensing Experiment"* *It is to be noted that OGLE has to due with:*

- *the optical lensing of Starlight, not Microwaves which is Totally unrelated to the subject matter of my paper.*
- *My paper focuses primarily on the Minimum-Energy or the Least-Time path of Microwaves Propagated in the Plasma Limb of the Sun and Sun-like stars. All observations and Experiments to date have reported that Microwaves deflect only at the plasma limb of the Sun and most importantly at the Lowest or the Least Impact Parameter.*
- *Microwaves are recorded to deflect only at 1.752 arsec, never at angles less than that or at Higher Impact Parameters. The 1/R Newton Potential has absolutely nothing to due with the effects described in my paper. The Impact Parameter is predicted by the Light Bending Rule of Relativity to be proportional to 1/R where R is the radius of the Analytical Gaussian Spheres inclosing the Solar Mass.*

Referee report

1. It does not contain new results. The Author uses a very well known light bending formula in the weak limit to evaluate the bending at different distances from the gravity center (different impact parameter). The Author illustrates these trivialities in Figures and in a Table

2. Author's divagations on the physical nature of the bending is all wrong - the bending is NOT due to the 1/R Newton potential (as he claims) but due to the curvature of spacetime.

3. Let "d" be the shortest distance of a light ray from a (spherical) mass and "R" the radius of the mass. The Author claims that the bending was observed only for d/R=1. This is not true. In the case of the Sun, the bending was observed at d/R=3 during several total eclipses. In the case of bending (microlensing) observed e.g. by the OGLE team, bending at d/R=800 is typically observed.

The paper should thus should be rejected.

*These are the Chosen Referee's own words.*

"The referees and the editorial staff of mainstream publishers are apparently hired to protect an in-house review system." -- Dr. Edward Dowdye

Fig. 5.25

Many people don't know that Arthur Eddington sent a second team to Brazil to observe the May 29th, 1919, solar eclipse. Eddington wanted to compare his data with the data collected by his second team in Brazil. But the newspapers were already printed declaring Einstein "proven right" overnight. When Eddington got word of the results from the Brazil team, he swept it under the rug and wouldn't have any of it. The values collected indicated that Newton was correct and the position of stars change during the eclipse due to refraction.

When stars change position during an eclipse, they do so ONLY at the boundary of the sun at 1 solar radius. Like how a straw changes its position at the surface (meniscus) of water. The position of the straw changes suddenly and without any distortion. The position of the straw changes ONLY at the boundary of the water due to a different refractive index. The same equation for the position of the stars changing under Lorentz invariance and relativity is the same equation for light deflecting under Galilean variance under Fermat. Relativity attributes the "bending" the gravity, and Newtonian/ Galilean mechanics attributes it to refraction. But refraction requires the presence of condensed matter which is forbidden and denied in the mainstream theory for the sun.

Semi-metallic hydrogen, type I metallic hydrogen and type II metallic hydrogen satisfy the requirement for the deflection of light due to refraction. But that wasn't known at the time of Eddington. Metallic hydrogen was first speculated around 1935, but that wasn't even a consideration for the composition of stars until recently.

The existence of metallic hydrogen is accepted in so called "gas giants" like Jupiter and Saturn, yet mainstream astrophysicists can't make the leap to apply their own logic to a larger celestial body with even more pressure. Gas giants technically don't exist since they contain condensed matter metallic hydrogen liquids, solids and exotic states of matter. The celestial bodies should more accurately be called "liquid giants" as Alexander Unzicker wrote in his book called "*The Liquid Sun - A Coming Revolution in Astrophysics*." The book details the work of Dr. Pierre Robitaille and documented science on condensed matter physics pertaining to the composition of the sun. The composition of stars greatly affects the interpretation of how light is deflected by the sun.

These images come from Dr. Pierre-Marie Robitaille's video on his YouTube channel, SkyScholar. The video is titled, "E.H. Dowdye, Can Stars BEND LIGHT? General Relativity and Gravity with Dr. Edward Dowdye". https://www.youtube.com/watch?v=B_ixkOI4k8c

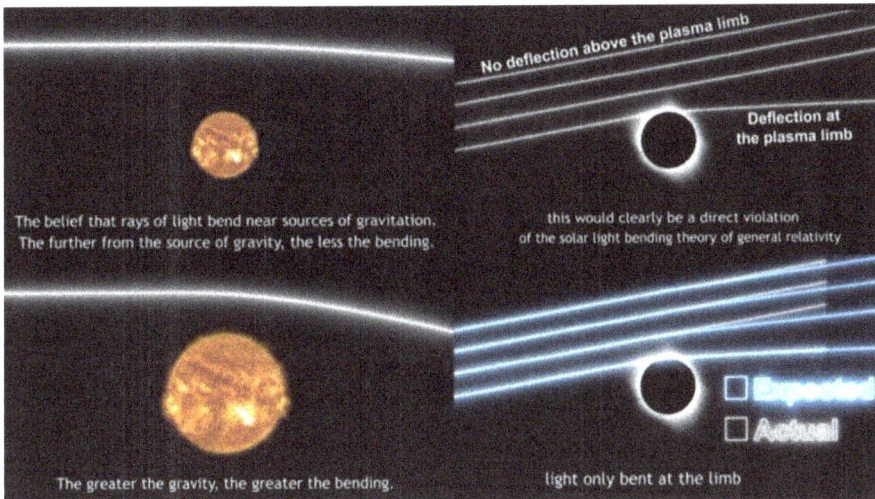

No deflection above the plasma limb

Deflection at the plasma limb

The belief that rays of light bend near sources of gravitation. The further from the source of gravity, the less the bending.

this would clearly be a direct violation of the solar light bending theory of general relativity

The greater the gravity, the greater the bending.

light only bent at the limb

☐ Expected
☐ Actual

Fig. 5.26

On the left is what Relativity and Eddington say we SHOULD see.... but on the right is what both Eddington and his 2nd team in Brazil actually observed on May 29, 1919. Refraction ONLY at the solar limb. Lorentz invariance invented all these equations to describe how gravity SHOULD bend light a given distance away from a body. There is a slightly different equation for each distance away at 1 solar radius, 2 solar radii, etc. equation for each distance.

According to relativity, the closer to the source of gravity, the more progressive bending of light following the inverse square. The further away from the source of gravity, the less bending following the inverse square. But light doesn't deflect according to the inverse square. It REFRACTS. (as per Fermat). Stars change their position WITHOUT distortion. Like a straw bending at the meniscus of water.

The straw does not progressively bend as it approaches the surface of the water. It just suddenly changes position AT the surface ONLY. Due to a different refractive index because of the presence of condensed matter.

Photo: Royal Society of London

One of Eddington's photographs of the May 29, 1919, solar eclipse. The photo was presented in his 1920 paper announcing the successful test of general relativity.

Fig. 5.27

You can look at the straw in the glass of water in infrared, visible light, x ray, microwave, radio, etc. You'll see the same thing in each wavelength and all spectra equally. We can see that with the shifting of the position of stars during a solar eclipse. That's validated all around the world. But again, the stars change position ONLY at 1 solar radius and without distortion. Relativity says there should be a progressive distortion of light due to it being stretched from following the curvature of space-time.

Gravity itself is defined as "the curvature of space-time" in relativity. And since mainstream solar theory treats the sun and stars as 100% balls of gas held together by "gravity"... they attribute any deflection of light to the curvature of space-time. All because they refuse to acknowledge the speed of light shifts. All these crazy interpretations of space-time distortions, gravitational lensing and literal time dilations are all due to a refusal to admit the speed of light shifts. All because of the illusion of electrons absorbing and re-emitting from their own frame of reference.

I just think it's ridiculous how a grainy black and white photo was quickly accepted as "proof" of relativity when Eddington's own team in Brazil argued against the data collected by the first team. The heat in Brazil slightly warped the housing of the 15 inch telescopes that were used, so Eddington discounted the totality of data... because it dismantled his interpretation of the data. That is confirmation bias because Eddington and Erwin Finlay were zealots. Here is an article from APS regarding Eddington's 2nd team:

https://aps.org/publications/apsnews/201605/physicshistory.cfm

Here are four quotes from the article:

"Comparison of the baseline measurements of a star's position and the corresponding measurements made during the eclipse, when that star was just visible at the limb of the sun, would determine whether Einstein or Newton was right."

"By the 1960s, most physicists accepted that Einstein's prediction of how much light would be deflected was the correct one."

"Not everyone immediately accepted the results. Some astronomers accused Eddington of manipulating his data because he threw out values obtained from the Brazilian team's warped telescopes, which gave results closer to the Newtonian value. Others questioned whether his images were of sufficient quality to make a definitive conclusion."

"When his assistant asked how he would have felt had the expedition failed, Einstein is said to have quipped, "Then I would feel sorry for the dear Lord. The theory is correct anyway.""

That goes to show that people blindly assumed the deflection was due to space-time curvature because the values were "close enough." And I take Einstein's statement as being an admission that his own cognitive dissonance was his guide. He would refuse to listen to reason if when presented with unreconcilable discrepancies. Einstein thought a grainy photo toppled the data from telescopes.

This equation from Dr. Dowdye represents the change in velocity of light/ electromagnetic radiation due to the change in the gravitational potential gradient. This derivation is yielded without need for a distorting space-time medium. Just as the deflection of light can be accounted for without need for a space-time medium. Just acknowledge the speed of light shifts. That's it.

$$\Delta v = v_0 \frac{GM}{Rc^2}$$

Galilean variance arrives at same predictions and solutions as Lorentz invariance. But without need to invoke a made up medium that's deviously conjured under a mathematical artifice. The world accepted the word of Eddington and a grainy photo as irrefutable evidence of space-time curvature. Without any third party analysis or considering his own team's data. I would take that as confirmation bias. Science is supposed to be about questioning oneself and not so much a journey to validate oneself by any means.

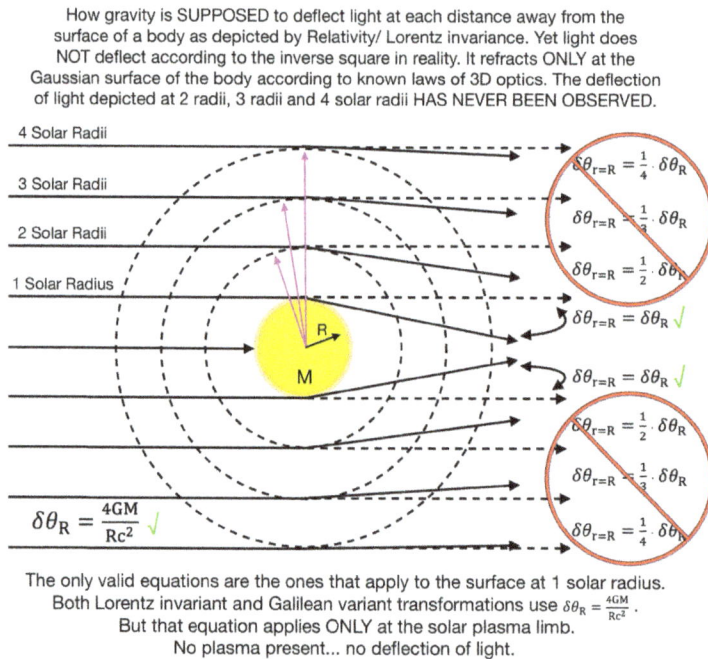

How gravity is SUPPOSED to deflect light at each distance away from the surface of a body as depicted by Relativity/ Lorentz invariance. Yet light does NOT deflect according to the inverse square in reality. It refracts ONLY at the Gaussian surface of the body according to known laws of 3D optics. The deflection of light depicted at 2 radii, 3 radii and 4 solar radii HAS NEVER BEEN OBSERVED.

The only valid equations are the ones that apply to the surface at 1 solar radius. Both Lorentz invariant and Galilean variant transformations use $\delta\theta_R = \frac{4GM}{Rc^2}$. But that equation applies ONLY at the solar plasma limb. No plasma present... no deflection of light.

Fig. 5.28

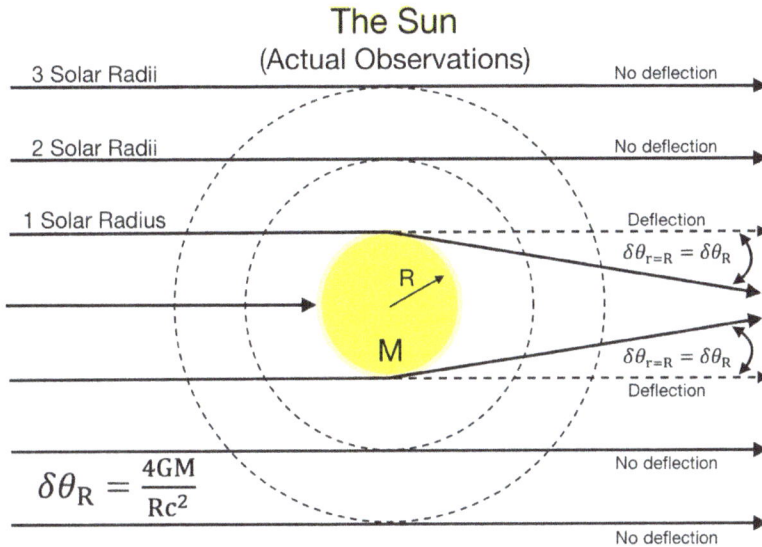

The Sun
(Actual Observations)

3 Solar Radii — No deflection

2 Solar Radii — No deflection

1 Solar Radius — Deflection
$$\delta\theta_{r=R} = \delta\theta_R$$

R

M

$$\delta\theta_{r=R} = \delta\theta_R$$
Deflection

No deflection

$$\delta\theta_R = \frac{4GM}{Rc^2}$$

No deflection

No deflection

Light follows a "least-time" path/ minimum-energy path/ mean free path.
Light does not deflect according to the inverse square.
Light deflects ONLY at the Gaussian surface, at 1 solar radius.

Fig. 5.29

This is an example of Dr. Edward Dowdye's findings on the deflection of light within a gravitational field. There is no deflection of light because of gravity. It's due solely to the index of refraction of the solar plasma limb. The gravitational potential gradient acts on the limb and the light is deflected within that thin plasma limb/ the solar atmosphere. There is semi-metallic and type I metallic hydrogen upwelled and ejected from the solar surface in flares. The metallic hydrogen is self-luminous in some cases. So there is plenty of justification for the deflection of light being due to refraction and having nothing to do with a non-Euclidean space-time medium which relies on an invariant model of light.

By simply acknowledging the presence of condensed matter and that the velocity of light shifts, we can do away with outdated nonsense about gravity bending light or any alleged lensing taking place. Any alleged lensing should drop off according to the inverse square, should be seen everywhere there is gravitation, even without plasma present, and should also be seen equally in all wavelengths and spectra. If there are *any* deviations from those relativistic consequences, it's devastating to Einstein's life's work. And indeed, there are plenty of examples.

We attack the legs of relativity by going after the illusion of the constancy of the velocity of light in all frames of reference, and then addressing the composition of the sun being condensed matter which justifies the total abandonment of gravitational lensing. And these findings are substantiated by observational and experimental evidence from the Max Planck Institute and more.

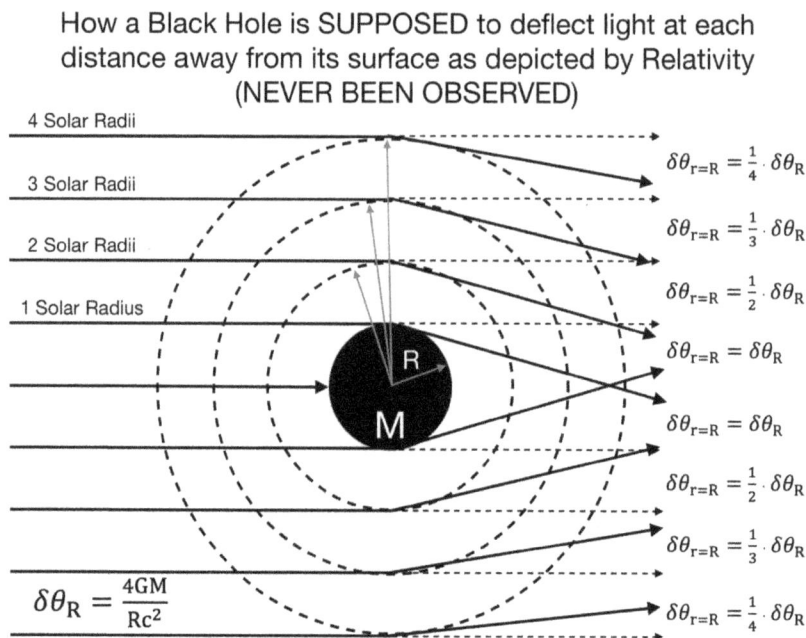

Fig. 5.30

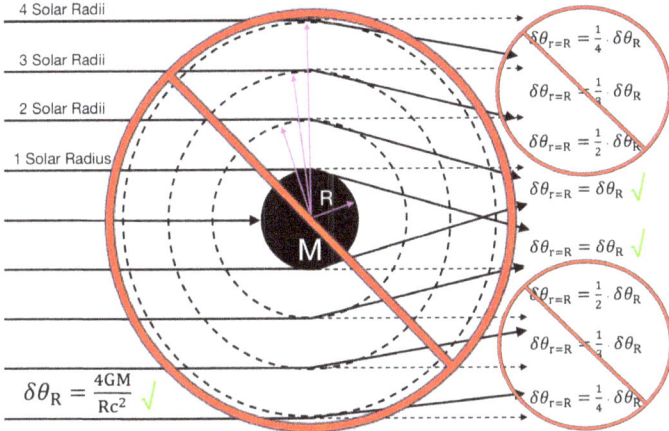

Only the equations dealing with deflection directly at the Gaussian surface of a body apply in reality. But there must be plasma present to refract/deflect the light. Relativity/ Lorentz invariance makes up a bunch of invalid equations for deflection at each radius. But not even the equations at the surface apply to Sagittarius A, because there is NO deflection at all. Not even directly at the alleged surface.

4 Solar Radii

3 Solar Radii

2 Solar Radii

1 Solar Radius

$\delta\theta_{r=R} = \frac{1}{4}\cdot\delta\theta_R$

$\delta\theta_{r=R} = \frac{1}{3}\cdot\delta\theta_R$

$\delta\theta_{r=R} = \frac{1}{2}\cdot\delta\theta_R$

$\delta\theta_{r=R} = \delta\theta_R$ √

$\delta\theta_{r=R} = \delta\theta_R$ √

$\delta\theta_{r=R} = \frac{1}{2}\cdot\delta\theta_R$

$\delta\theta_{r=R} = \frac{1}{3}\cdot\delta\theta_R$

$\delta\theta_{r=R} = \frac{1}{4}\cdot\delta\theta_R$

$\delta\theta_R = \frac{4GM}{Rc^2}$ √

Lorentz invariant equations are plugged into computer simulations for black holes which they think justifies the outcome of lensing for their simulations. Faulty equations = faulty computer graphics. Garbage In Garbage Out (GIGO)

Fig. 5.31

A celestial body like the sun of a given mass has a specific radius at the boundary/surface of that body, (the Gaussian surface). According to relativity, the deflection of light is supposed to take place abiding by the inverse square, following the drop off of gravity away from the surface of a body. If the sun had mass, but didn't have an ionized plasma atmosphere, there would be no means for light to deflect.

The mass of a body alone SHOULD "bend"/deflect light according to relativity, but if there's no plasma present to refract the light, then there is no deflection at all. When it comes to an alleged black hole... there is no "surface". Gravity is defined as "the curvature of space-time itself" in relativity. There is no gravitational force under the framework of Lorentz invariance That model refers to the gravity dropping off from the core (as does Newtonian mechanics), but in Newtonian mechanics, we deal with tangible real-world surfaces in 3D Euclidean space. But the core of an alleged black hole is said to be a "singularity"... which is said to have "infinite mass with zero volume." Infinite weight with no size at all they say and having no surface. But there is no space-time under Galilean variance. Black holes are claimed to produce "infinite space-time curvature" under Lorentz invariance. And that erroneous take is based upon the assumption that the velocity of light remains the same in all frames of reference. There quickly begins a slippery slope of points to address due to the logical fallacies of relativity. Black holes do not exist in Galilean variance.

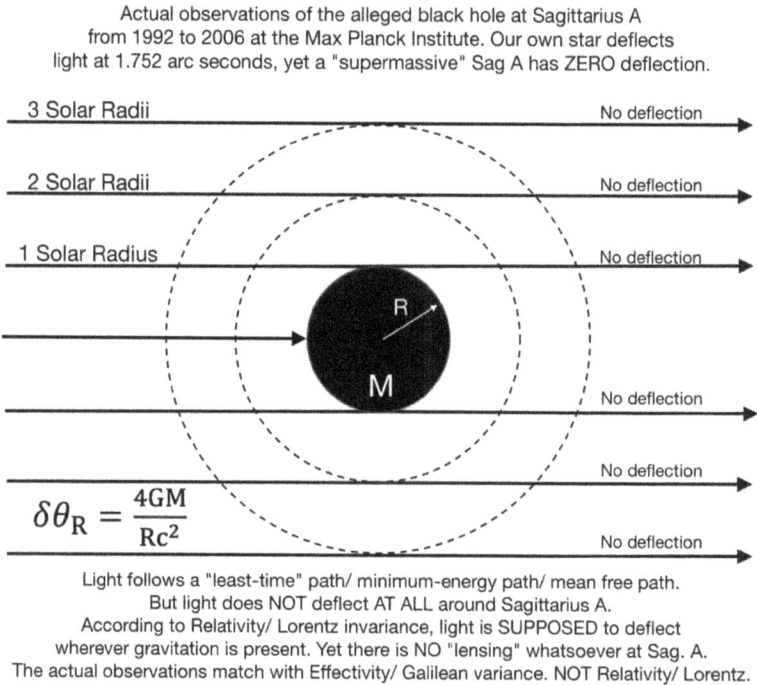

Actual observations of the alleged black hole at Sagittarius A
from 1992 to 2006 at the Max Planck Institute. Our own star deflects
light at 1.752 arc seconds, yet a "supermassive" Sag A has ZERO deflection.

$$\delta\theta_R = \frac{4GM}{Rc^2}$$

Light follows a "least-time" path/ minimum-energy path/ mean free path.
But light does NOT deflect AT ALL around Sagittarius A.
According to Relativity/ Lorentz invariance, light is SUPPOSED to deflect
wherever gravitation is present. Yet there is NO "lensing" whatsoever at Sag. A.
The actual observations match with Effectivity/ Galilean variance. NOT Relativity/ Lorentz.

Fig. 5.32

So it's not the mass directly that actually deflects light... it's the plasma atmosphere AROUND the mass. And a mass in reality has a definitive radius and surface. But the deflection of light requires an atmosphere of plasma around the condensed matter body. The radius of the Gaussian surface will determine the angle of deflection using what Dr. Dowdye called the "plasma focal length". That is denoted as the Greek letter, $\xi$ .

$$\alpha = \frac{4GM}{\xi c^2}$$

A given mass has a given radius. Only when that mass has a plasma atmosphere present does that atmosphere have an index of refraction and act like a lens to deflect light, and ONLY at the Gaussian surface of that mass. Sagittarius A does not have a plasma atmosphere. An alleged black hole doesn't have a surface regardless of its supposed mass and there is no deflection of light to be found at or around Sagittarius A.

One cannot deny the apparent motion of stars around the location of Sagittarius A with the evidence collected through time resolved imagery. But the reasoning for why the stars maintain their orbits under Lorentz invariance/ relativity is that it MUST be a black hole and/or dark matter. Yet we are left with the discrepancy that our own local star with an average density of $1.41 \frac{g}{cm^3}$ deflects light at 1.752 arc seconds and yet an alleged "supermassive" black hole with at least 20 solar masses more than Earth's Sun doesn't deflect light AT ALL.

The mass required to form a mythical black hole is 20 times greater than the mass of our own local Sol star. And since there is much more gravity/ alleged space-time curvature claimed for a black hole... then you should figure that since our own Sun deflects light at 1.752 arc seconds... that a supermassive black hole should certainly deflect light at least 20 times greater than that... right? Yet there is ZERO deflection of light whatsoever at Sagittarius A. There goes the argument that gravity bends light. Sagittarius A has more gravity than our own Sun yet doesn't produce any lensing whatsoever. Not even at what they claim is supposed to be the Gaussian "surface".

The reason for the lack of lensing at Sagittarius A is unknown. I reject the model of black holes in totality. At this time, I have two explanations for the lack of lensing at Sagittarius A. One is that there is a dense body present made of condensed matter. No black hole, no curving of space-time, no gravity bending light. Just a dense body, probably made of exotic matter like metallic hydrogen, metallic helium, metallic neon and other exotic states of matter that only exist under extreme pressures (or manufactured coherence).

The model of gravitational collapse is a misnomer since it's based on a star being 100% gas/plasma. As James Jeans suggested, (cite paper), if a star contains any condensed matter at all... gravitational collapse becomes impossible and stars are limited to maximal densities and volumes and real factors in 3D Euclidean space.

Sun on Trial - Dr. Robitaille: https://youtu.be/9TOKo7Ik9f8?si=3wLJSW9XZi6i0jKe

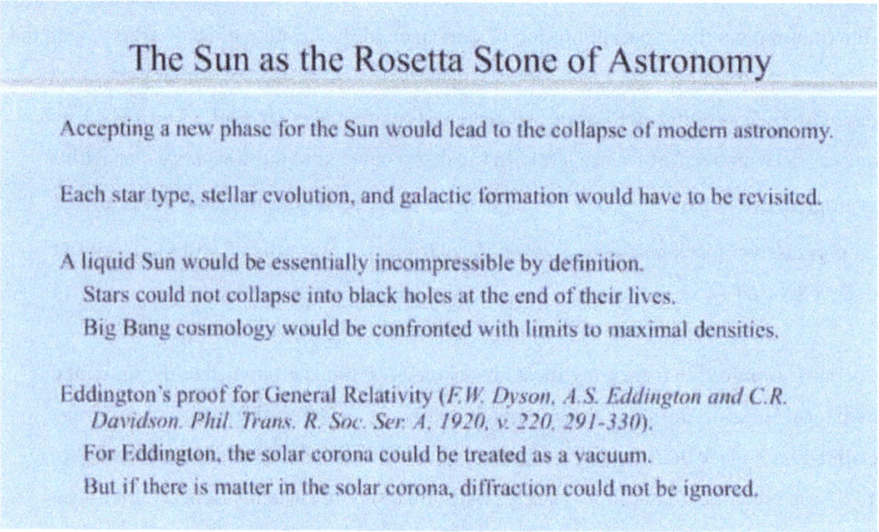

## The Sun as the Rosetta Stone of Astronomy

Accepting a new phase for the Sun would lead to the collapse of modern astronomy.

Each star type, stellar evolution, and galactic formation would have to be revisited.

A liquid Sun would be essentially incompressible by definition.
  Stars could not collapse into black holes at the end of their lives.
  Big Bang cosmology would be confronted with limits to maximal densities.

Eddington's proof for General Relativity (*F.W. Dyson, A.S. Eddington and C.R. Davidson. Phil. Trans. R. Soc. Ser. A, 1920, v. 220, 291-330*).
  For Eddington, the solar corona could be treated as a vacuum.
  But if there is matter in the solar corona, diffraction could not be ignored.

Fig. 5.33

The model of gravitational collapse suggests the ENTIRE "body" violates thermodynamics by the system doing work on itself.. raising its own temperature while decreasing its radius... all down to a mathematical fictional singularity of "infinite density and zero volume". Infinite weight with no size at all they say. But if there is just a dense metallic body incapable of emitting light since it's behaving like an optical blackbody... then the gravitational collapse might only apply to the *atmosphere around* the condensed matter celestial body. So the dense body is incapable of retaining an atmosphere as the gravity is too great. No gases can just linger in an atmosphere. Just boundary of the Gaussian surface of the condensed matter body and the void.

Since gravity has no effect on light... and the ONLY deflection of light observed in space is at the Gaussian surface of a body with plasma present at 1 solar radius... and if a body that dense is incapable of retaining an atmosphere... then there would be no deflection of light at all even though it's massive.

The second explanation is that there is no dense body there at all. It _used_ to be there. Gravity is not some mythical curvature of space-time. Gravity _is_ a force and an emission that propagates like light. The light and gravity that already left a star... that light and gravity will continue to propagate out even if that star extinguishes or is destroyed. Play a speaker. Now turn off the speaker or destroy the speaker. The sound that already left will continue to propagate out and cause an echo even if the speaker is destroyed. The same applies to the gravity that leaves a star.

Earth is not orbiting the exact center of the Sun. We are orbiting were the center of the Sun _used to be,_ 8.33 minutes ago. The gravity that left the center of the star travels in a straight line. But by the time that part of the wavepacket reaches us... the Sun has moved its position in the sky by 20.5 arc seconds. So the light and gravity Earth receives from the Sun is 8.33 minutes old. If the Sun just vanished like Thanos snapping his fingers... Earth would still continue to orbit where the center of the Sun _used to be_ for another 8.33 minutes. We would have perfect weather and light for 8.33 minutes until a wall of darkness approached at the speed of light from the direction of where the Sun was. Then the back-end of the gravitational wave packet and light from the sun would suddenly cease.

Back to Sagittarius A... if there was a dense body there at one time emitting gravity and maintaining the orbit of the surrounding stars... and the gravity emitted by that body has a delay to reach the surrounding bodies at the velocity light, c.... and that central star extinguished or exploded... then the surrounding bodies would continue to orbit where Sagittarius A _used to be_ for however long it takes the light and gravity to propagate at c from the central star to those bodies. But, like a plasmoid as discovered by Dr. Winston Bostick... the surrounding orbiting bodies could establish a self-contained loop based upon the already established orbits. And given that the gravity from those stars would be re-emitting each other's graviton packets, the orbit of those bodies might be sustained for some time like an echo, node or some non-linear function.

The deflection of light from the Sun is different than the gravitational lensing claimed for distant galaxies. The observations of distant objects are not the same in different spectra or range of frequencies. Relativity claims the observations should be the same throughout all spectra. The alleged lensing effect claimed by relativity does not discriminate between spectral ranges. The streaks and arcs seen in the ultraviolet should also be seen in infrared, but they are not. I'm going to pick an alleged example of lensing touted by NASA called ABEL 2218.

There should be no variations at all between spectral bands. It should be totally independent of frequency according to relativity. But every example of alleged lensing is frequency dependent. The same features attributed to gravitational lensing just do not appear in other frequencies. Direct violation of relativity. Some people might point to the Herschel wavelengths of ABEL 2218 But those aren't claimed to be features of lensing. The lensing is claimed to be far *around* the object itself.

https://newscenter.lbl.gov/2010/01/19/weak-lensing-gains-strength/

Image credit: ESA/Herschel/SPIRE & NASA/ESA/STScI

The image above shows the Herschel image (centre) in relation to an iconic image of Abell 2218 from the Hubble Space Telescope (right). The three images on the left show the three Herschel wavelengths independently. The centre of the cluster is marked as a white cross-hair, and the bright yellow object just below is the lensed galaxy.

Fig. 5.34

The "example" of strong lensing shown with visible distortion is ONLY captured in that isolated frequency and only when the light is amplified like turning the brights on your car in the fog. The forward light scatter produces streaks and ONLY in that frequency.

The "statistical averaging" is more devious use of isolated bands to try and show the lensing exists... but it doesn't exist. It's flat out fraud because gross negligence is too innocent to overlook something of this magnitude when you have near unlimited resources for 100 years. If the distortions aren't equal in all frequencies, then gravitational lensing doesn't exist. If the distortions don't show up in time resolved imagery, then gravitational lensing doesn't exist. If there's no deflection of light where there is gravity like Sagittarius A, then gravitational lensing doesn't exist. Drop the relativistic nonsense please.

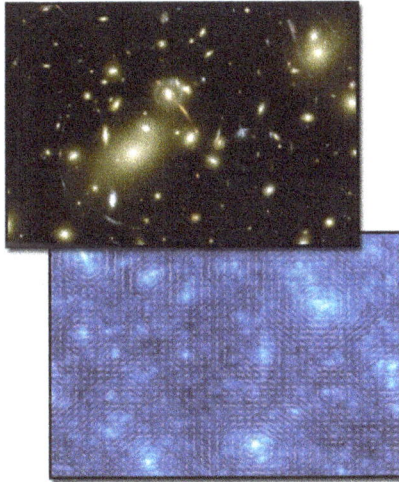

A spectacular example of strong gravitational lensing is the nearby galaxy cluster Abell 2218 (top), in which the visible distortion of individual background galaxies can be used to measure the mass of the lensing structure. The weak lensing of fainter and more distant structures must be detected by statistical averaging (bottom). (Abell 2218 image by NASA, weak lensing simulation by Bhuvnesh Jain, Uroš Seljak, and Simon White)

Fig. 5.35

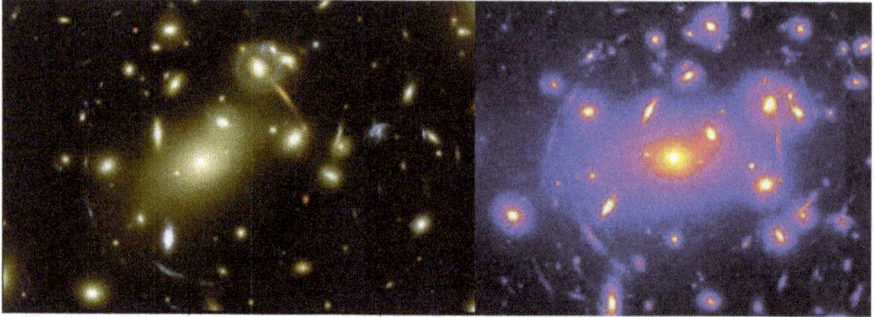

On the left - Abell 2218  NASA. A. Fruchter & The ERO Team (STScl) • STScl-PRC00-08 (HST • WFPC2)   Image captured in 1999

Fig. 5.36

On the right – "Abell 2218 Cluster is a photograph by W.Couch & R.Ellis NASA uploaded on May 10th, 2013."

Fig. 5.37

The image on the right from 2013 is just the same image from 1999 but rotated clockwise about 53 degrees. I overlayed the two images and it's the same image just with false color. It's then pawned off as if the same ABEL 2218 but taken at two different times or at the same time with two different spectral filters. But it's not. The original image was taken in the near infrared. The claim for the 2013 image is that it's in the *microwave*. You can't retroactively "reformat" data for one spectra and just apply it to another band to generate the same image in another color as if the image itself was captured in that frequency band.

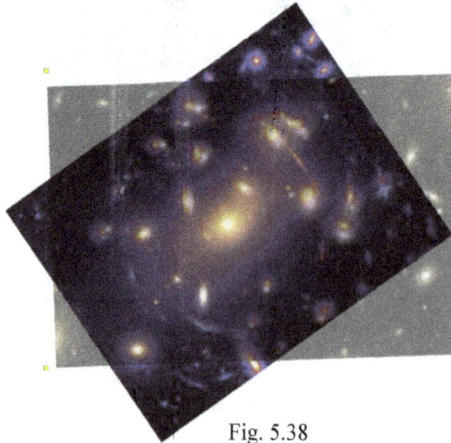

Fig. 5.38

That's devious and deceitful to me, but other people in the industry apparently just call that "reprocessing." By that logic, Einstein "reprocessed" the aether and overlaid space-time on top of it then called it a new "photograph." The claim is that the image is a "visible light image" Ok... so wait... let's go over this one more time... The 1999 ABEL 2218 image was captured in NIR, broad visible light (only red and blue) and UV... The image shown for the 1999 ABEL 2218 image was only presented in near infrared (NIR). The data for the other broad visible light and UV was just sat on until 2013. W. Couch and R. Ellis "reprocessed the image"... which is claimed to be "a visible light image".

Here's what they did to fool people. It's a bunch of overlaid images of multiple layers. It's a composite image. The original image data from 1999 showed an artifact of streaks and arcs. Those streaks and arcs are *only* in the NIR. There are no streaks and arcs in the broad visible or UV bands. But they kept the data for the streaks and arcs in the NIR and then overlaid the blue and red on top of that. (which when overlaid again makes the purple).

So, to maintain the underhanded streaks and arcs, they left that part of the NIR data in while only extracting and overlaying the blue and red to overlay on top of the NIR data. They layer the original to make it appear as if the streaks and arcs also appear in the broad visible blue and red. Then they rotated the image under the guise of "aesthetics."

There is no side by side image of the claimed lensing effects of *any* given location between different spectral bands. There is *no* time resolved imagery of *any* lensing effect as claimed by relativity, there's *zero* comparisons between multiple spectra of the same location for a given lensing claim, none of that. That would require the scientific method. Remember as Michio Kaku said, "No one I know uses the scientific method. It's by the seat of your pants! Leaps of logic." (These dudes have been playing leapfrog with each other a bit too much...)

Different color mapping was used from the F450W and F606W filters. That's what they consider "visible light" but it's really just an isolated red and blue. There is no "photograph" that's a misleading term to use. They just took the data from the same NASA source and overload the rendered images from all that data on top of each other. Layer upon layer.

NASA Corporate needs you to find the difference between

This Picture    and    This Picture

Fig. 5.39

"They're the same picture"

Fig. 5.40

This is the case with every alleged image of lensing. Every single one of them can be attributed to forward light scatter and illusions of using a flawed tool. Or at least the tool is used nefariously to amplify the light many times brighter until the "lensing" effect appears. That so called lensing, again, only happens in isolated spectra after amplifying the light way more intense than is naturally. Gravity does not affect light. Gravitational lensing doesn't exist. Asking for "time resolved imagery" of the lensing effect claimed by relativity will be met excuses pouring from frothing mouths. The relativist's credibility relies on lensing being uniform throughout all spectra, with or without amplification. The onus is on anyone in the entire astrophysics community to provide just one example of the lensing effect IN MOTION. Almost two decades of observation from the Max Planck institute showed no lensing whatsoever at Sagittarius A yet cartons and textbooks still depict fantasies that have never been observe red. It's a giant cult and a racket.

Just like William de Sitter's double star experiment is refuted because any wave packets are extinction shifted by gas clouds, dust and media that causes forward light scatter. It's like turning on your brights in the fog. The fog (medium) causes glare and lens flair and forward light scatter. Sometimes the glare in space only shows up in certain wavelengths and frequency bands. The bent light in all the images people point to on Wikipedia and NASA is artifacts.

An artifact in imaging can happen in astronomy, spectroscopy, medical imaging and all types of imaging. Artifacts refer to distortions or errors or flaws in the image that aren't really there. That's really important in X-rays for patients so they don't get misdiagnosed with having a growth or not. An artifact can show up in an X-ray giving the false impression a mass is present when it might not be. It could just be an artifact of the X-ray image due to many various factors.

An artifact in astrophysics isn't really talked about as much. Every image peddled by NASA is assumed to be meticulously reviewed for any discrepancies. Yet the biggest discrepancy of all is that each claimed artifact or distortion should be seen EQUALLY in ALL spectra, and in motion (time-resolved imagery). Never been observed. Show me, don't tell me.

Here is a contradiction with a lack of lensing produced by spiral galaxy pair NGC 3314. This image won the Hubble Heritage Program Photography Award for release number STScl-2000-14 with the link here :

https://web.archive.org/web/20120901221757/http://hubblesite.org/newscenter/archive/releases/2000/14/image/a/format/web_print

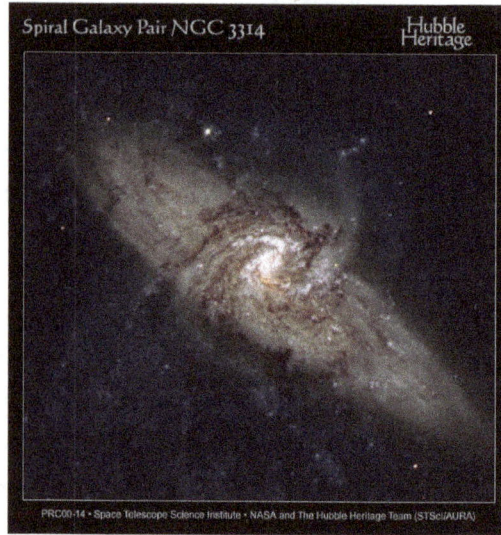

Fig. 5.41

According to relativity, the galaxy in the foreground should produce lensing and bend the light to distort the galaxy in the background. The two galaxies are in a rare line-of-sight alignment. The foreground galaxy eclipses the background galaxy. There should be light bending, rings, arcs and distortions galore, but there are none... zero... zilch.

All the images of alleged "Einstein crosses" like QSO 2237+0305, HE 0435-1223, RX J1131-1231, WFI J2033-4723, HE 2149-2745 and PG 1115+080 are misconceived as well. The spectral similarity between the "split up stars" in each Einstein cross is attributed to redshift. But since redshift is simple a difference in relative velocity independent of acceleration or direction, the different stars in the Einstein cross are just that... different stars. Not the same star split up because of lensing. It's all just assumptions by relativity.

Another example of an apparent contradiction of prediction versus reality is the G2 gas cloud. A computer model was rendered to predict what *should* happen to the G2 gas cloud as it passed near an alleged black hole from 2012 to 2016. But actual time-resolved imagery taken over that period shows the G2 gas cloud was mostly unaffected by the alleged gravity that should have ripped it apart and attracted it to the black hole. There are plenty of examples showing a contradiction of what *should* be versus what *is*.

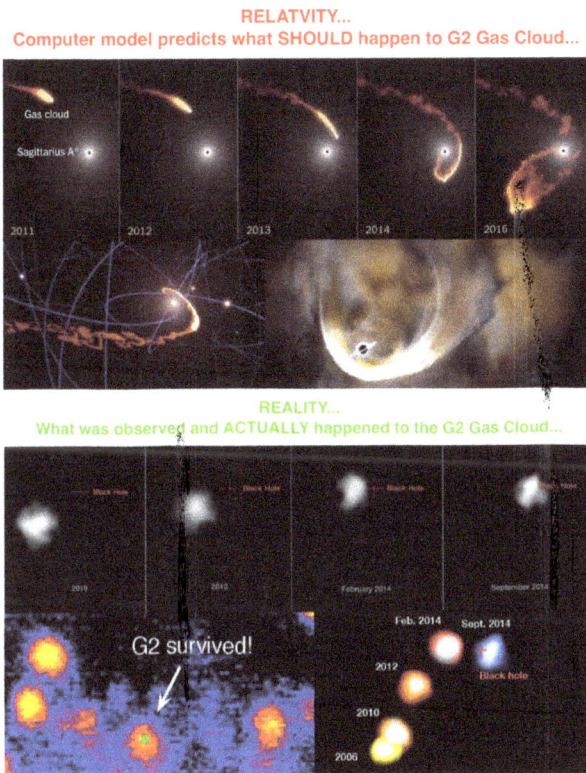

Fig. 5.42

How can an alleged mysterious body trap light with its own gravity to the point where light can't escape, yet the gravity of that very body can't even deflect light more than our own star? And the same black holes which are all alleged unescapable bodies (which isn't a body at all) doesn't even affect a "mere" gas cloud. The contradictions are too great to reconcile.

There are many more unreconcilable discrepancies. Relativity says gravity is not a force at all and then invokes space-time curvature to influence the path of light and imposes non-Euclidean rules on 3D Euclidean space. Under Galilean variance however, there is no space-time and we deal with reality in good ol' 3D Euclidean space. No infinities, no paradoxes and gravity is viewed as a force.

Isaac Newton just sat up in his coffin like the Undertaker from the WWE hearing a church bell.

Fig. 5.43

Challenges to Gravitational Lensing and More:

https://web.archive.org/web/20210907060452/http://www.extinctionshift.com/Signific
antFindings.htm

Gauss' Law for gravity and observation reveal no solar lensing in empty vacuum space:

https://vixra.org/pdf/1807.0149v1.pdf

Gravitational Light Bending History is Severely Impact-Parameter Dependent

http://www.naturalphilosophy.org/pdf/abstracts/abstracts_6523.pdf

Gravitational Lensing in Empty Vacuum Space Does Not Take Place:

https://www.naturalphilosophy.org/pdf/abstracts/abstracts_5973.pdf

The Shapiro Delay: A Frequency Dependent Transit-Time Effect

https://www.naturalphilosophy.org/pdf/abstracts/abstracts_6105.pdf

E.H. Dowdye Jr., 2018, The Solar Plasma Limb is found to deflect Microwaves from Extra
Galactic Radio Sources at Lowest Impact Parameter Corresponding to the Solar Plasma Limb

https://web.archive.org/web/20220124082433/http://www.extinctionshift.com/PaperS
ubmittedToAstronomischeNachrichten.pdf

"Time resolved images from the center of the Galaxy appear to counter General Relativity",
Dowdye, Jr., E.H., Astronomische Nachrichten, Volume 328, Issue 2, Date: February 2007,
Pages: 186-191. Published on-line at:

http://www3.interscience.wiley.com/search/allsearch
(Search under author: Dowdye)

INFINITE ENERGY * November/December 2009 * Issue 88
"Are the Conventional Concepts of Gravitational Lensing Adhering to the Observational
Evidence and Mathematical Physic Fundamentals?", Infinite Energy, Vol 15, Issue 88, 2009

http://www.infinite-energy.com/iemagazine/issue88/index.html

# Magnetism Doesn't Directly Affect Light

You can clearly see the light bending here in the Crookes Tube, right? No one would dare deny the apparent empirical evidence and objective observation of light bending... right?

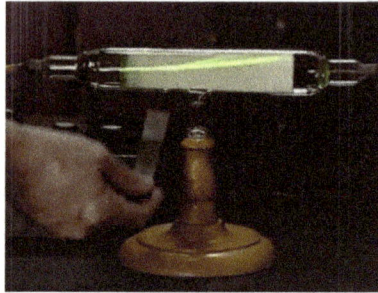

Fig. 5.44

No, it's illusions of observation. We see the path of charged particles being *traced.* The magnetic field changes the path of the glowing particles. The particles flow in an arc. It is Not the light that bends. Light does not bend. It is re-emitted/ deflected at the boundary of a medium. The deflection of light is sudden and occurs at a tangent, at a constant. A brand-new light is re-emitted at an angle within a medium due to an index of refraction.

But in the Crooke's tube example, the opposing magnetic field is oriented 90° to the current of charged particles. The charged particles want to continue flowing in a straight, collimated beam, but the magnetic field influences them 90° away from their straight-lined path; therefore, the particles flow in an arc. But the light emitted by those charged particles is something different than the particles themselves. The light merely exposes the path of the particles. The alleged "bending of light" in the Crooke's tube is only INDIRECT.
A produces B. C affects A which indirectly affects B. Electrons produce light. Magnets affect electrons which indirectly affects the light. The same experiment with lasers doesn't yield bending.

Now that we've gained more perspective on how light propagates... let's apply that to gravitation and compare to other models of gravity.

— CHAPTER 6 —

# MODELS OF GRAVITY

Continuing the theme about light, shadows and gravity,

***All the gravity we ever feel is the re-emitted gravity from the electrons making up ourselves.***

When you talk about "gravity" with people, what specifically do you mean?
When asked about gravity, one must give an accurate definition based upon the consequences of a given model in order to do an answer justice. It's known what gravity does. No one is denying that when you jump up, you come back down. But it's the mechanism of how that is not understood. The explanations for the mechanism of gravity are contradictory and incomplete when it comes to relativity's interpretation. No one denies the apparent difference in clocks, but it's the mechanism that is not properly understood which leads to erroneous interpretations.

Here are some factors that people confuse with gravity:

- Electrostatic Attraction/Repulsion
- Magnetism
- Eddy Currents
- Centrifugal force
- Acceleration
- Casimir Effect
- Coanda Effect/ VTOL
- Gyroscopes
- Buoyancy
- Equalizing of Pressure
- Acoustic/Ultrasonic Levitation
- Optical Trapping
- Ion Wind
- Meissner Effect/ Flux Pinning

All these factors *SIMULATE* the *EFFECTS OF* gravity... but none of those things *ARE* gravity.

None of those factors listed above are an even spherical influence towards the core of a body. Isaac Newton never claimed to have a theory for the *cause* of gravity. He developed the mathematical proofs to describe the evidence of the behavior of matter and energy within a gravitational field and a gravitational potential gradient.

When you refer to "gravity", do you mean Newton's model which says gravity is a force? That mass attracts mass (without any electric influence)?
"I have not as yet been able to discover the reason for these properties of gravity from phenomena, and I do not form any hypotheses..." -- Sir Isaac Newton
https://en.m.wikipedia.org/wiki/Newton%27s_law_of_universal_gravitation

Fig. 6.1

Albert Einstein said gravity is *not* a force. Relativity and Lorentz invariance define gravity as the curvature of space-time itself. The deformation of a 4th dimensional medium that simultaneously has and does not have properties like a condensed matter medium. There is no attraction, no pull, and no work exerted on anything whatsoever according to the model of gravity under relativity. Like the ol' bowling ball on a mattress visualization. The curve in the mattress does not exert *any* force on anything else on, above or around the mattress.

Remember the *Toothpick Bridge Analogy* from the beginning of the book. This is another example of relativity taking aspects of classical physics to use as glue for the holes in its own theory. Every depiction of space-time curvature attempts to convey their idea using physics reserved for condensed matter. "Bending" is a term reserved for the deformation of condensed matter and 3D topology. The entire notion of space-time curvature is egregious nonsense. If your mind has trouble visualizing something, you may not be suffering from aphantasia. If you can't imagine the scenario, that might be because the scenario isn't how reality operates. Gravity as an emission is simpler than relativity's space-time curvature.

- Einstein said gravity is NOT a force and defined gravity as the curvature of space-time itself. The deformation of a 4th dimensional medium.
- There is no gravitational pull or attraction in Relativity. It is inappropriate for supporters of relativity to use the phrase "gravitational force" or model any pull.
- Gravity is the displacement of a medium according to Relativity. (Like sitting in a bathtub. A 3D mass displaces a 4D space-time)
- The action/reaction of displacement of a medium is SIMULTANEOUS/ INSTANT.
- In condensed matter, the effects of displacement throughout a volume of a medium does NOT propagate at the rate of C, nor do the effects of displacement drop off according to the inverse square from a mass displacing that medium. Displacement ripples, like capillary waves, can accelerate outward at the surface of condensed matter.
- Forces propagate & real-world factors in 3D space drop off according to the inverse square.
- Both gravity and luminosity drop off according to the inverse square away from a light source, but time does not drop off according to the inverse square. Time is supposed to be linked with light and gravity in relativity. But it's not in reality.

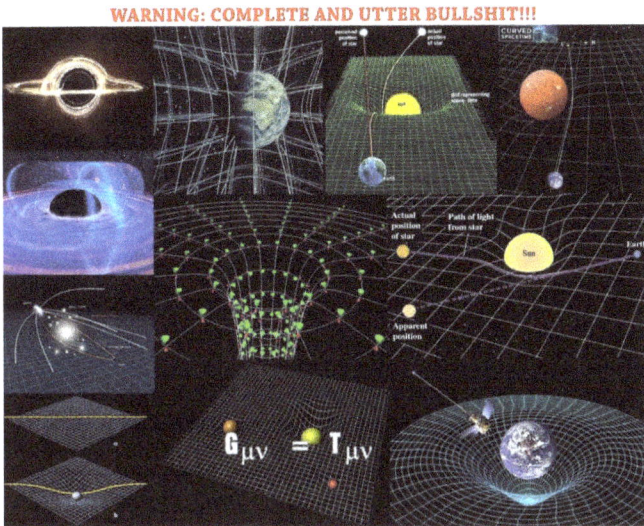

Fig. 6.2

Fig,. 6.3

Dr. Dowdye's *Extinction Shift Principle* and Galilean variance challenges general relativity's Lorentz invariance without invoking space-time curvature or distortions in time.

***All the light we ever see is the re-emitted light from the electrons making up ourselves.***

The photons that leave a source are not the same photons interpreted by an observer. Likewise, *all the gravity we feel is the re-emitted gravity from the electrons making up ourselves.* It is never "the same" gravity that propagates from a source and affects a body or observer. A primary gravity is absorbed, and a secondary gravity is re-emitted from the electrons making up whatever absorbed that primary. And that brand-new secondary gravity propagates relative to those electrons. The gravitons that leave a source are not the same gravitons affecting a body. Those packets can be in phase or out of phase.

Assuming gravity propagates like light under Galilean variance, that would mean a brand-new gravity is produced at a constant by a source, like a star. Then, yet more new emissions are re-produced at a constant by the electrons making up whatever blocked and absorbed the primaries. But the re-emissions are generated from incoherent fermions and not electron pairs or bosons like an initial primary source. This serves the illusion that gravitons don't exist when measured. People are looking for literal ball particles or direct evidence of primaries, when neither photons nor gravitons are particles at all, and direct measurement of any primary emission is physically impossible.

All the gravity capable of being measured is the re-emitted secondary gravity from the electrons making up the measuring devices or interferometers or observers themselves. We can measure the effects of gravity but cannot directly measure the primary influence or cause of gravity. Therefore, it is physically impossible to detect any incoming primary emissions of gravity from any source. The number of electrons making up an observer or measuring device is simply too infinitesimal and negligible and incoherent to detect the results they desire because 1) it is assumed an observer sees and feels the same light and same gravity from a source. And 2) it's not possible to directly measure any primary.

People got stuck on the illusion of a distorting medium and abandoned the true nature of light and gravity. The primary gravity from a source might be generated specifically by a system in a coherent state. And the electrons making up any target are not necessarily in a coherent state, so the re-emissions are different than the primary. But there's no physical way to measure that at this point. Because, again, the electrons making up any measuring device would absorb the incoming primaries and re-emit secondaries relative to itself. The act of measurement extinguishes that which you are trying to measure. Thus, the complete dismissal of gravity being an emission.

The refusal to acknowledge the velocity of light shifts has led to a lot of delusions from illusions. By acknowledging light speed shifts, it does away with any need for linking space with time and for that linked space-time to behave like the surface of water. A relativist dismissing the accurate predictions and solutions of Galilean variance is like a flat earther dismissing geodesy. Solving the illusions of re-emissions explains the alleged "missing matter" or lack of baryonic matter that was invented to justify the amount of gravity observed in the cosmos.

Another illusion is the difference between coherent gravity from a coherent source verses the re-emitted incoherent gravity from an incoherent source. The re-emitted gravity from incoherent electrons prevents any physical measurements of incoming primary gravitons from coherent sources. The plane wavefronts are extinguished upon attempting to measure them, which then produces spherical wavefronts.

What other than light can duplicate the properties of illumination?
There isn't anything. Light IS illumination. A photon is the emission and light/illumination is the result of absorbing and re-emitting that primary influence.

What other than gravity can duplicate the property of a constant spherically symmetric downward force towards the core of Earth?
There isn't anything. Gravity IS gravitation. A graviton is the emission and gravity/attraction is the result of absorbing and re-emitting that primary influence.

# What is Light? What is a Photon?

Light is photons. Illumination is the absorption and re-emission of photons. In an ideal void (so-called vacuum), a photon/ wavepacket is a massless emission propagating in a straight-lined path from a source, but outward in all directions forming a total sphere. A brand-new energy is generated at a constant and projected out from a point source at the rate of c according to Huygens' Principle. The difference in energy between the electrons making up the source and the electrons making up the target will determine the frequency and wavelengths of light. And the speed the source is traveling towards or away from a target will determine that light's velocity relative to each other. Light is not "a wave" traveling between planets or atoms. Photons are not made of particles despite the claims made for the double slit experiment. Photons are not "made of" anything. Being "made of something" means to have mass and contain matter. All electromagnetic radiation is massless.

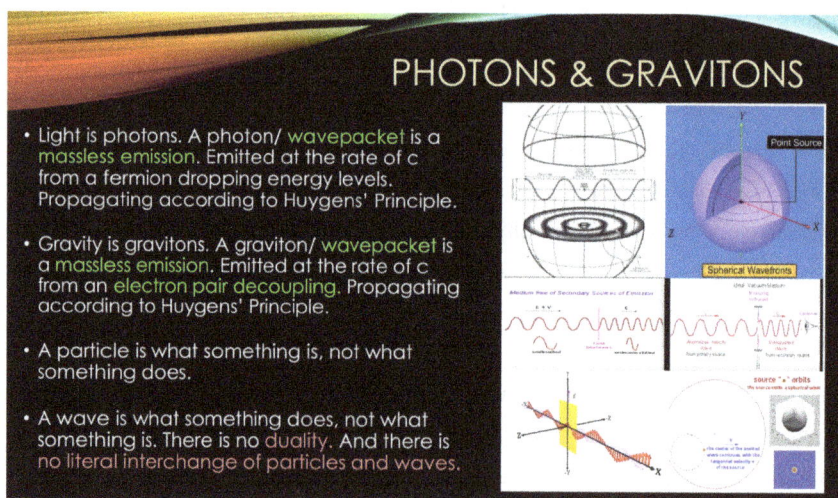

Fig. 6.4

Here is a reference to a peer reviewed article from Dr. Edward Henry Dowdye Jr. "The photon and its measurability", Proc. SPIE 5866, The Nature of Light: What Is a Photon? (4 August 2005); https://doi.org/10.1117/12.621454

# What is Gravity? What is a Graviton?

Gravitation is the result of absorbing and re-emitting gravitons. Just as illumination is the result of absorbing and re-emitting photons. The mechanism of how gravity is emitted is different than that of light. But the premise is similar. Both gravitons and photons are the result of a lowering of energy states. Both light and gravity are emitted, absorbed and re-emitted by electrons. A gravitational field is established when an initial coherent plane wavefront is emitted from the decoupling of electron pairs at a constant rate. In an ideal void (so-called vacuum), a graviton/ wavepacket is a massless emission propagating in a straight-lined path from its source at the rate of c, but outward in all directions forming a total sphere. A brand-new energy is generated at a constant and projected out from a point source at the rate of c according to Huygens' Principle. It's not the same light or gravity at any point or instant within the spherical boundary of an emitted wavepacket. A wavepacket is not "1 object" but that's how light is treated. All light is the same according to Einstein.

The difference in energy between the coherent electrons making up a source versus the Brownian motion of electrons making up a target will determine the frequency and wavelengths of gravity between the two masses. Additionally, the speed the source is traveling towards or away from a target will determine that gravity's velocity relative to each other. Gravity is not "a wave" traveling between planets or some mythical curvature of space-time. (I address LIGO in Book 2 of the series on Gravity.)

Gravitons are not "made of" particles. I've heard the saying attributed to Richard Feynman stating, "*a wave is what something does, not what something is.*" I think that statement can be expanded upon. Given there is no particle-wave duality in Galilean variance, the other side of the coin is, "*a particle is what something is, not what something does.*" The ocean is not made of waves. The ocean is made of water, and water can wave. Light is not "made of" waves. Light is not "made of" any "thing." The so called "wave" people refer to is generated at the point of interference when blocking/absorbing an incoming primary packet of constant influence. The rate (frequency) of absorption and emission of electrons is affected.

Electrons are affected in a repeated pulsation as an equal and opposite reaction. The target blocking the path of those primary emissions will experience the frequency of interference between the electrons making up itself and the source. This applies to both light and gravity.

## Emission & Re-emission of Gravity

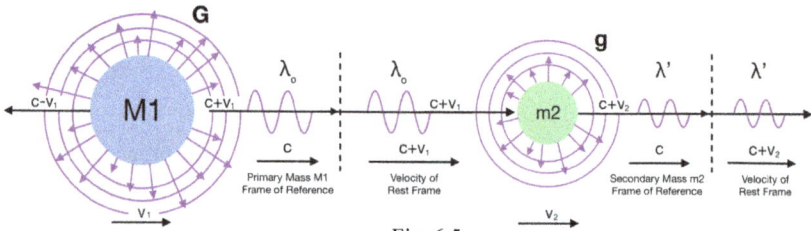

Fig. 6.5

The primary gravitational wavepacket from the larger source, M1, propagates with velocity C relative to M1. M1 itself travels at velocity $V_1$ relative to that mass' frame of reference.

The primary wavepacket propagates with wavelength $\lambda_o$ and velocity $c + v$ relative to that rest frame. It would have the velocity $c + v_1 - v_2$ relative to the smaller mass, m2 that re-emits and mutually influences M1. The back and forth influence of gravitational emissions and re-emissions cause the smaller mass to oscillate or jiggle at the frequency:

$$v' = \frac{c + v_1 - v_2}{\lambda_o} = v_0 \left(1 + \frac{v_1 - v_2}{c}\right)$$

Once established, the gravitational field re-emitted by the smaller mass m2 reciprocally influences the larger mass M1 with a secondary gravitational packet at the wavelength of :

$$\lambda' = \frac{c}{v'} = \lambda_0 \left(1 + \frac{v_1 - v_2}{c}\right)^{-1}$$

That wavelength is propagating at the velocity c relative to the smaller mass m2.

That wavelength is simultaneously propagating at the velocity of $c + v_2$ relative to the rest frame and velocity $c + v_1 - v_2$ relative to the larger mass M1.

For more mathematical details, refer to Dr. Dowdye's book, ISBN: **0-9634471-5-7**

Here are a few more equations from Dr. Dowdye that are apropos. These come from his presentation he gave at CNPS July 20th and 21st, 2016 with John Chappell. The name of the presentation was called "Microwaves from Extra Galactic Radio Sources found to Deflect only at Minimum Impact Parameter Corresponding to Solar Plasma Limb." The equations appear at the 40 minute mark to 41:30. https://youtu.be/-6X-vv97zEE?si=uSTWsjGWjPmARxal

Integrated Effect of the Gravitational Gradient Field (Potential) of mass M:

$$\Phi_{r=R}^{r=\infty} = \int_{r=R}^{r=\infty} \frac{GM}{r^2} dr = \frac{GM}{R}$$

This represents potential due to a mass M, but in the context of the deflection of light, it's analogous to the refractive index variation. A given mass has a given radius. When there is a plasma atmosphere present to deflect the light, the medium causes the re-emission according to the gravitational potential gradient as listed above. This also denotes the distance to the mass and the radius of curvature of the body. The deflection is dependent upon the medium with is dependent upon the mass of a given radius. That affects the re-emissions of light at the Gaussian surface of the mass.

Max Velocity equation. This represents the escape velocity in the context of gravitation.

$$V = \sqrt{\frac{2GM}{R}}$$

Redshift of a Frequency of light:

$$v' = v_o \left(1 - \frac{v^2}{c^2}\right) = v_o \left(1 - \frac{2GM}{Rc^2}\right)$$

Redshift of a Wavelength of light:

$$\lambda' = \lambda_o \left(1 - \frac{v^2}{c^2}\right) - 1 = \lambda_o \left(1 - \frac{2GM}{Rc^2}\right)^{-1}$$

$$\therefore \lambda' \approx \lambda_o \left(1 + \frac{2GM}{Rc^2}\right)$$

Number of Wavelengths per unit Length:

$$n = \frac{1}{\lambda'} = \frac{1}{\lambda_o \left(1 - \frac{v^2}{c^2}\right)^{-1}} = \frac{1}{\lambda_o}\left(1 - \frac{2GM}{Rc^2}\right)$$

The shifts in wavelengths occur in a refractive medium. The above equation represents changes due to variations in the density and ionization of the plasma atmosphere surrounding the mass of a given radius.

The Energy along a path of light as a function of Gravitational Potential:

$$\varepsilon = \varepsilon_o\left(1 - \frac{2GM}{rc^2}\right)$$

Change in Energy per unit length:

$$\frac{d\varepsilon}{dr} = +\varepsilon_o\frac{2GM}{r^2c^2} \qquad \delta\varepsilon = +\varepsilon_o\frac{2GM}{r^2c^2}\delta R$$

Changes in energy can be due to variations in the refractive index of the media in addition to (or instead of) gravitational potential. The above equation accounts for the refractive index of the atmosphere or medium by which the light is deflected.

The angle of deflection upon approach:

$$\delta\theta_{app} = \frac{\delta\varepsilon_{app}}{\varepsilon} = +\int_{r=\infty}^{r=R}\frac{2GM}{r^2c^2}dr = -\frac{2GM}{Rc^2}$$

The angle of deflection upon receding away:

$$\delta\theta_{rec} = \frac{\delta\varepsilon_{rec}}{\varepsilon} = +\int_{r=R}^{r=\infty}\frac{2GM}{r^2c^2}dr = +\frac{2GM}{Rc^2}$$

Net Deflection Angle of light at the Gaussian surface (1 solar radius):

$$\delta\theta = \delta\theta_{rec} - \delta\theta_{app} = \frac{4GM}{Rc^2}$$

The angles of deflection under Galilean variance are due to refraction in a plasma medium.

Another important understanding of gravity comes from the work of Siméon Denis Poisson. Poisson (pronounced "pweh-saw") wrote his equations using Newton's universal law of gravitation. No space-time curvature, tensors or 4th dimension nonsense. Easy and concise.

| Aspect | Standard Poisson Equation for Gravity | Galilean Variant with Coherent Potential |
|---|---|---|
| Equation | $\nabla^2 \Phi = 4\pi G \rho$ | $\nabla^2 \Phi_{\text{coherent}} = 4\pi G (\rho_{\text{coherent}})$ |
| Interpretation | The standard equation relates the gravitational potential $\Phi$ to the mass density $\rho$ in space. | The modified equation includes a coherent potential $\Phi_{\text{coherent}}$, incorporating the effects of coherent electron pairs and external forces (e.g., Lorentz force), and relates it to the coherent mass/energy density $\rho_{\text{coherent}}$. |
| Source of Gravitational Field | Mass density $\rho$ produces a gravitational field and potential $\Phi$. | Coherent mass/energy density $\rho_{\text{coherent}}$ from orthogonally bound electron pairs generates the potential $\Phi_{\text{coherent}}$, which can modify the gravitational interaction through coherent energy storage and release. |
| Potential Nature | The potential $\Phi$ is a scalar field describing gravitational attraction due to mass distribution. | The coherent potential $\Phi_{\text{coherent}}$ is a scalar field representing energy stored in a coherent system, like orthogonal electron pairs in a 2D plane, influenced by external forces such as the Lorentz force and released longitudinally upon destabilization. |
| Energy Release Mechanism | There is no explicit mechanism for energy release in the standard Poisson equation. | The coherent potential describes a system where stored potential energy is converted to kinetic energy upon release. The system may maintain coherence until destabilization, at which point energy is emitted, potentially leading to spherical wavefronts. |
| Boundaries and Emission | In a standard gravitational system, the mass distribution determines the boundary conditions and field propagation. | The coherent system's Gaussian surface represents the boundary where the coherent potential transitions to kinetic energy. This boundary may be spherical or asymmetric, depending on the system, and the energy emission could follow spherical wavefronts beyond the coherent boundary. |
| Scalar or Vector Potential | The gravitational potential $\Phi$ is scalar, representing energy per unit mass. | The coherent potential $\Phi_{\text{coherent}}$ is also scalar but includes the contribution from the external forces acting on the system (e.g., Lorentz force), giving it a unique dynamic in terms of energy storage and release. |
| Force Generation | The gravitational force $\mathbf{F} = -\nabla\Phi$ is derived from the gradient of the gravitational potential. | The force from the coherent potential is also derived from $\mathbf{F}_{\text{coherent}} = -\nabla\Phi_{\text{coherent}}$, but includes contributions from the orthogonal electron pairs' stored energy, which may lead to different force dynamics at the boundary of the coherent system. |

Fig. 6.6

# The Gravitational Spectrum

• Gravity has a spectrum like light. A primary gravity can be absorbed, and a secondary gravity re-emitted like a reflection/shadow/echo.

• Gravity has wavelengths and frequencies and propagates like light at the rate of c relative to its source.

• Gravity has a delay/offset and aberration effect.

• Gravity can be focused like a spotlight or even a LASER.

• Gravity IS a force!!

• All the light you ever see is the re-emitted light from the electrons making up yourself.

• All the gravity you ever experience is the re-emitted gravity from the electrons making up yourself.

• By this logic, there is Newtonian Gravity and "Goethenian" Gravity. When they merge, then create an entire spectrum which abides by the same laws of optics as light.

You don't have to worry about the gravity of the Earth, or the Sun, or the weight of the world Atlas... Ultimately, you only need to worry about electrons making up yourself.

Liberating Atlas: "Finally! The 'weight' is over"

Fig. 6.7

• Gravity is an emission propagating at the rate of c. *(Gravity IS a force!)*

Gravity is emitted, absorbed and re-emitted by electrons in coherent and incoherent states. Gravity is the result from a coherent source producing electron pairs at a constant rate which then split and decouple outside that coherent boundary. Gravitons are emitted at a constant rate as a result which are then absorbed at a constant rate by the electrons making up surrounding matter. The absorption and re-emission of those graviton packets are what we experience as gravity.

• Attractive force is generated from ratios of coherence vs incoherence and a differential of states that are in phase. A disparity of a larger mass with so much pressure that it generates a coherent state at the core. That state emits bosons at a constant which then split at a constant as the energy exist the boundary of that high pressure/ordered state.

• Repulsive force is generated from ratios of coherence vs coherence or incoherence vs incoherence and out of phase. Production of gravity from a coherent source will emit coherent gravity. If two coherent sources are emitting coherent gravity, they will repel. The mass of both bodies is never exactly equal in nature. Not even any two sides of a given object have the same amount of atoms and are not symmetrical. And the coherent gravity from the coherent core will generate one type of gravity while the matter outside that coherent boundary will generate another type of gravity that overlaps the first type. So there is always an offset of mass and gravity in nature. An equilibrium is met of attractive and repulsive forces to make for a stable and constant gap between bodies in orbit.

• The disordered electrons making up a smaller mass will be attracted to the ordered bosons of a larger mass. The incoherent energy rushes to fill the void of the ordered state in an attempt to corrupt it, Entropy has its way with an ordered state to create a flow in a particular direction until states are equalized over time. The coherent state need not do any work since the energy is already available in its potential.. The rest of the surrounding incoherent electrons will seek out the larger coherent mass, abiding by the 2nd Law of Thermodynamics.

• The fermions act LIKE a sponge and act LIKE a magnet or AS IF static cling. That attractive force is from a constant differential being produced between the 2 bodies on the subatomic scale, but in totality as macroscale objects with definitive physical boundaries, finite masses and specific radii.

• Light is not a force unto itself, yet light can impart momentum and act on matter like "radiation pressure" (which requires the presence of condensed matter) and the photoelectric effect (which is mainly in the UV range). Light is not a force but can act as a carrier to impart a force. The same must apply to gravitons under the assumption and consequences of the model. Gravitons are not a force, but the re-emission of gravitons as gravity IS a force that is mutually and reciprocally interchanged, absorbed and re-emitted.

• Gravitons are the carrier wave by which the gravitational force is imparted. Once the energy of the graviton is absorbed and re-emitted by the electrons making up something else... then gravity is the action resulting from the equalizing of potentials. As the frequency of interference between coherent source and incoherent receiver (if attracting). Likewise, the frequency of interference between coherent sources or incoherent sources being out of phase if repelling.

• The decoupling of electron pairs at a constant rate is the cause of gravity. The massless gravitons are the carrier and the action/feeling of gravity itself the result of absorbing and emitting the difference in energy between your own electrons and that gravitational source. That interaction happens at a constant no matter what part of the planet you're on because the majority of Earth is made of incoherent fermions, but the core is coherent bosons. And the incoherent electrons of Earth are attracted to the coherent electrons of the Sun. Scale invariant from atom to galaxy. We are constantly absorbing the gravitons from the Sun and Earth and surrounding matter. Then re-emitting it which attempts to equalize between the conflicting states resulting in things sticking to a celestial body.

• The same Light and Gravity does not propagate from the Sun and affect us on Earth. By the time it reaches Earth…. the primaries are absorbed and re-emitted by the Earth itself... and then subsequently re-emitted by all the electrons making up YOU.

• ANY photons in the visible spectrum produce the property of illumination. But just because a certain light might be out of the visible range for humans doesn't mean photons aren't present.  ANY gravitons will produce the property and Attraction and/ or repulsion between coherent and incoherent masses. Two like coherent states will repel. But there can be an equilibrium of short-range repulsion and long-range attraction. (Unless one body has enough momentum and relative velocity to overpower the repulsion between the two bodies and they collide.)

• More atoms = more electrons which give a false impression the mass itself is the cause. Electrons are electric in nature. Electrostatic potential with a gradient. NOT orbiting particles as depicted in textbooks using the Rutherford-Bohr model of the atom. I like a recent publication by curtis-press.com called "*The Nature of the Atom - Introduction to the Structured Atom Model*". The book is written by J.E. Kaal, A. Otte, J.A. Sorensen and J.G. Emming. You can check out more regarding a different model of the atom on the Structured Atom Model website: https://structuredatom.org

• Light is emitted by electrons. Yet light itself is not electric. The same applies to gravity. Gravity is not electric. Yet the high voltage discharge of exotic craft, UAPs, UFOs might serve the illusion that the gravity itself is electric. It is Not.

• There is NO direct link between gravitation and electromagnetism. ONLY INDIRECT. Thus, the futile search for a theory of everything or a unification of mutually exclusive concepts with opposing mathematical frameworks like Relativity and QED. Or unifying Classical physics with Relativity. Galilean invariance and Lorentz invariance conflict. Galilean invariance and Galilean variance conflict. Lorentz invariance and Galilean variance conflict.

At some point you've just got to pick a side and stop with the superposition of all possibilities all at once. Not everyone is right. Some things are definitively wrong and some people just suck.

Like repels like in regard to polarities. The North poles of two magnets will oppose and repel each other. The South poles of two magnets will oppose and repel each other. Same with electrostatics and static electricity. Two negatively charged objects will repel each other. Two positively charged objects will repel each other.

But in regard to states of density, opposites repel each other. Serving the illusion that same densities attract within a gravitational field. The density of a helium ballon is less than the atmosphere around it at sea level. Therefore, the atmosphere pushes the helium balloon up and away. The atmosphere rushes to fill the void of density which results in lesser density being pushed out of the way to make room for the denser medium.

The helium balloon will continue to rise until the buoyancy of the rubber balloon itself matches the pressure and density of the atmosphere at a specific altitude. But it will find a happy place to just chill out after a certain time. (but the balloon expands as it rises until the rubber stretches so much it pops the balloon eventually.)

But if the balloon never popped and there was no wind blowing the balloon around the helium gas it would just find a spot to rest in the atmosphere to equalize it's pressure and density potential and "specific gravity." The mass of the rubber of the balloon wants to fall back to Earth from gravity, but the Helium wants to continually rise, so the balloon with helium finds a "happy medium"... within the medium of the atmosphere.

Hang a pendulum in your car verses a helium balloon in your car.
As you accelerate forward, the pendulum which is heavier than the air around it will swing to the back of the car as you accelerate forward. But the helium balloon will push to the front of the car towards the windshield because the heavier air around it is sloshing to the back. The helium balloon does no work to move forward. It's the opposing denser medium

surrounding it that pushes it out of the way. That is not gravity though. You get buoyancy in a medium due to the surrounding gravitational field. But the surrounding medium with determine if you rise, sink or remain dependent upon your specific gravity and density in relation to the surrounding medium. Without a medium present however... there's no buoyancy to alleviate the effects of gravity at a given size. The medium is affected by gravity and the medium can stave off the effects of gravity while you're in a dense medium.... but the buoyancy within the medium is not representative of gravity.

Like polarities repel each other... but density and pressure and surrounding medium will determine how an object "floats" or remains suspended or moves to equalize that pressure. AS IF the like states of density or pressure attract each other. The helium balloon does not rise because it's attracted to the sky. A bubble does not rise to the surface of water because it is attracted to the surface. The heavier atmosphere and medium around the bubble or balloon pushes it upward and repels it away from that pressurized state. The pressure being generated from the atmosphere pushing off the condensed matter surface of the Earth itself and the ground. 2nd Law of thermodynamics again of equalization.

But what about coherent states? Coherent states can both attract and repel each other depending on the circumstance. Something in a coherent state will invoke the surrounding incoherence to corrupt it through entropy. Generating a flow until the incoherence and coherence is equalized. And the internal kinetic energy produced by the decoupling of electron pairs would repel that like-kind of energy. Therefore... generating a state of coherence intensely enough would compete with the coherence produced by the center of the Earth and wouldn't "lose weight"... but rather the Earth's gravitational field would be suspending the like-coherent mass with all of its weight. Serving the illusion of weightlessness or buoyancy. The repulsion between the Earth and an artificially induced coherent state would be equal to the difference of coherent potential being dissipated at a constant rate and that coherent potential can be at a given frequency and wavelength when it discharges as a kinetic coherent current. Those coherent currents can have a constant decoupling of electron pairs if that state surpasses a critical threshold of superconductivity.

Other forces and effects confused as gravity:

With electrostatic cling... if you're clinging... you aren't able to freely walk around. You are bound to a point like fly paper until more force is exerted than the amount of force of the cling. Or like an electrostatic bell as developed by Andrew Gordon, a monk from around the 1740's. The electrode of the electrostatic bell neutralizes upon touching the surface and then reverses.

Magnetism doesn't attract non-ferrous solid materials. And you notice that a magnet is not free to slide or move about on a steel surface. The effect drops off at the equator of the magnet and is not spherical. Diamagnetic materials like water, blood, copper are repelled by eddy currents which are produced in opposition to the magnetic fields. Pyrolytic graphite is an extremely diamagnetic material that hovers over the joined corners of magnets that attract together naturally, (rather than forcing them together.) Magnetism and eddy currents are not gravity though.

Casimir Effect can *simulate* attraction like gravity, but it's not spherical attraction. It's between two physical boundaries which cancel out the frequencies correlating with the distance between the boundaries. So, the Casimir plates or boundaries are pushed together from the outside. Ships on water that are docked too close together are pushed together from the waves on the opposing sides of the ships. The space between ships can cancel the ocean waves between the ships, resulting in the ships crashing together from a fluid mechanical analog of the Casimir effect. But the Casimir effect isn't gravity.

Centrifugal force is produced equatorially. If a medium of liquid is spinning inside a closed chamber, the centrifugal force will push the medium outward and fling it away from the blades. The high energy liquid shoots off the blades and moves the rest of the medium out of its way. As an equal and opposite reaction, the liquid medium is pushed against the walls of the chamber which then loop back around and push down. As the medium is energized by the blades of a blender let's say, the total cycle of pushing out and pushing in serves the illusion of a pull force. A vortex forms. But the vortex is ultimately all the result of a push.

The centripetal spin of the blades initiates a centrifugal force as a reaction. There is no centripetal force. Centripetal is an action of a spin. Torque is the force but the centripetal spin describes the action of spinning in place like a basketball on your finger. The spinning in place forms the pushing force at the equator of the spinning object. That indirectly forms the vortex in a closed chamber. But a medium pushing off the walls of a chamber is not an equally spherical force toward a core. Centripetal spin and centrifugal force is not gravity.

Inertia is in the opposite direction as acceleration. But inertia can be staved off as you accelerate up to a certain threshold given you have the coherent potential in available electron pairs at a given moment to oppose the inertial forces. It's not the acceleration directly but rather the inertia resulting from the acceleration which people take as being equal to gravity. But that too is not a spherical influence. Acceleration is in one direction. Gravity is felt an equally spherical attraction to a core. Relativity wrongfully thinks centrifugal force, acceleration and gravity are one in the same. Far from it. Acceleration is not gravity.

Buoyancy can simulate levitation depending on the mass of the object in relation to the surrounding medium. (The relative densities.) There is a "happy medium" of objects floating in a given density media in the regard they will stay suspended in the air or water. The effect can be likened more to inverse gravity more so than a constantly spherical downward force. Buoyancy results in objects pushing *away* from the surface, not attracting to it. Buoyancy is not gravity.

The Coanda effect or magnus effect works for an object in opposition to a current of air, and that spinning object is fighting gravity, so it, "floats".

Acoustic and Ultrasonic levitation is not spherically attractive. Acoustic levitation requires a cubical chamber with three speakers oriented 90 degrees to each other in three dimensions. X, Y and Z axis. In an experiment uploaded to YouTube by Dr. James A. Deak.
https://youtu.be/94KzmB2bI7s?si=1e7F9mLjPbP0UwCo

There are three horns acting as waveguides for the 600 hertz tone leading from the speakers into the Helmholtz resonant cavity (plexiglass box). The sound pressure creates a focal point and node or point of lesser sound compared to the rest of the inside of the box. The object in the box is pushed to the point of least resistance. But the mass of the object being lifted is only a ping bong ball or a dry porous sponge or Styrofoam, etc. If you wanted to lift something with more mass... the amount of sound pressure required would crumble and dis-integrate the object apart before it lifted. And the amount of sound/ decibel level would deafen anything for miles. By adjusting the amplitude and the phase of the three speakers, you can manipulate the focal point to move in 3D space and/or spin on its axis in either direction. Acoustic levitation is not gravity or anti-gravity. Poor Dr. Deak had to slay misnomer after misnomer in the comments of his video until he finally got fed up and shut down the comment section. Too many people who don't read the pinned post and run with their ideas about Stonehenge being levitated into place and how the pyramids were built from the top down. I've got your back James.

With Ultrasonic levitation, it only requires two transducers oriented 180 degrees apart. Similar result of a focal point of deconstructive interference, or node, created by the intersecting wavepackets. A single plastic bead or drop of water or something small can be levitated within that node. The effect of both acoustic and ultrasonic levitation is not a continuous and spherical influence toward a core. Interesting though, the water droplets and small particles are pushed away from the antinodes and focal points of constructive interference. So apply that to actual emissions of gravity from decoupling electron pairs and we have something more as an analog. Ultrasonic levitation itself is not gravity though. But out of all the examples so far, the ultrasonic transducers is *most like* a gravitational laser. But not exactly gravity still.

Ion wind is directional and requires condensed matter to push against, whether that be the surface of the earth itself. But, again, that is inversed to gravity and away from the ground, not attracting towards the core. A large mass will push away a smaller mass. A smaller mass will bounce off of a larger mass. Kinetic energy will be transferred but the smaller object goes flying off more than the bigger object. A large object ejecting a bunch of tiny objects can create thrust. But if the ejections of the large object are too small, then there will be no thrust.

When people are looking for "anti" gravity, they are mostly talking about directional scenarios or pushing AWAY from a particular surface. Not constantly falling towards a core.

Same with the Meissner effect. The Meissner effect is the actual lifting and pushing away from the surface of a material subjected to critically low temperatures. Like a small magnet pushing up and away from the surface of a superconductor due to the magnetic field being blocked/ expelled from the superconductor. If you weigh the superconductor and magnet.... then pour liquid nitrogen and see the magnet lift... you will see the scale stays the same weight because the magnet's mass is being pushed up and away from the superconductor with all its weight added even though it's not touching.  And the effect is not spherical TOWARDS a core.

So... just like how there is nothing else other than the production of photons to produce the properties of illumination... there is nothing else like the production of gravitons to produce the spherical mystery of "attracting" to the Earth's surface as if trying to continually fall to the core.

The frequency shift of that gravity from primary to secondary and source to absorber generates that attractive pull.
It's a dimensional shift from pent up 2D energy and releasing back to 3D. The graviTONS are generated and the absorption and shift of frequencies from the re-emission of graviton energy results in the attraction we call graviTY.

So, Newtonian gravity and Einstein's Relativity directly link mass to gravity. And things eventually fall apart for both models. Newton said it's a mass dependent force that attracts other mass somehow. And Einstein said, no, no... it's a mythical curvature of space-time, not a force at all, but that gravity is indeed mass dependent. Both Newton and Einstein were incorrect to try and directly link gravity to mass. Newton had the mechanics to account for the effects of gravity but never claimed to explain a cause for gravity. Einstein attempted to explain the cause for gravity but still had to use Newtonian mechanics to explain the effects. By correctly identifying the mechanism of gravity, we can rid ourselves of the fallacious argument that A directly affects C. There is no direct link between mass and gravity. Only Indirectly linked. There is no direct link between gravitational and electricity. Only indirectly linked. There is no direct link between the so called "quantum" scale (subatomic) and the macro scale when it comes to gravity... only indirectly linked. Attempting to directly bridge them is futile and destined to oblivion.

Newton got it righter than Einstein though for sure. Only thing Einstein got right really was the stimulated and spontaneous emission. But that's optics and electric engineering in 3D Euclidean space which applies Newtonian mechanics. And we can return to the correct foundation of Newtonian mechanics with the additional understanding of a Galilean variant framework for light. That framework for light directly relates to gravitation. And the mechanism for the production of gravity is the foundation of electron pairing and decoupling within a coherent state.

It's when the electron pair formation outpaces the capability for the pairs to split at an even rate that we get the production and over-production of gravitons. When there's more energy coming in 3 dimensions and only so much able to output in 2 dimensions, a dimensional phase transition occurs to produce gravity, emissions of light, negative ions and other unforeseen anomalies.

# Tending the Garden of Science and Knowledge

Science is a journey and a process to gain knowledge through experience. Placing unnecessary hurdles or obfuscation into a scientific process hinders the evolution of mankind. The gardener is not the garden. The painter is not the painting. The composer is not the composition. But the gardener, painter and composer all have a great respect for that to which they tend.

If someone came into the garden and poured salt or oil on the soil, that would piss off the gardener for putting the crops in danger and all future crops in danger. If someone came into the art studio and swapped out all the fine hair bristle brushes with stainless steel bristle brushes, the painter would be pissed off for putting the canvas in danger and all future art projects in danger. If someone came into the music studio and made sure every string of the pianos in the building were the same length and invariant... the composer would be pissed off for making his masterpiece futile to play and put all future compositions in danger. Like... how dare you!

Lorentz invariance is like pouring salt and oil on the soil, like replacing all the brushes with steel bristles and like making all the strings of the piano the same length. It's a DEFCON 1 danger to humanity's future to correctly interpret nature's processes in reality. People have no clue that the invention of the fourth dimension and space-time is so reprehensible and irresponsibly dangerous for the future of mankind. How are the gas prices in your area again? Wait... we still use gas? ...Wow.... Well, the good news is, I hear nuclear fusion is only 50 years away. (Again/still) Enjoying your relativity based world which states the propositions in this book should be discarded and banned like YouTube did?

Relativists are vandals who (un)knowingly participate in the destruction of their own world and/or prevent their own world from flourishing. All because they refuse to stop worshipping the golden calf they call Albert Einstein and his invisible friend, the fourth dimension. Lorentz invariance ruined the world. It's time to start tending to the garden.

# List of Figures & Image Credits

## Chapter 1

Fig. 1.1 - (Page 4): 22nd Century Science logo
*Image Credit: Author's Own*

Fig. 1.2 - (Page 6): Isaac NEWTon
*Image Credit: Author's Own*

Fig. 1.3 - (Page 8): Einstein | Dowdye split screen
*Image Credit: Author's Own*

Fig. 1.4 - (Page 10): MOGGED Theory
*Image Credit: Author's Own*

Fig. 1.5 - (Page 10): Einstein divorce court/ball n' chain
*Image Credit: Author's Own*

Fig. 1.6 - (Page 11): Multiple images
(a) Spiral clock (top left)
*Image Credit: Courtesy of Brighton Science Festival 13 February - 6 March 2011*
(b) Black hole image from the movie Interstellar (bottom left)
*Image Credit: The supermassive black hole Gargantua from the 2014 sci-fi blockbuster "Interstellar." Paramount Pictures*
(c) Interstellar movie poster (left)
*Image Credit: Movie Poster for Interstellar (2014). Directed by Christopher Nolan. Warner Bros. Pictures. Artwork by Drew Struzan*
(d) Double clock (right)
*Image Credit: Double clock located at the Times Building in downtown Indianapolis, Indiana. Photographer unknown*
(e) Sun refraction only at the solar limb (top right)
*Image Credit: Slide number 67 from Dr. Dowdye's presentation on the Extinction Shift Principle, uploaded by Peregrine O'Brien'.*
*https://slideplayer.com/slide/7361784/*
(f) Shifu Yan Lei meditating (bottom right)
*Image Credit: Instant Health: The Shaolin QiGong Workout For Longevity, Shifu Yan Lei*
*https://www.amazon.com/Instant-Health-Shaolin-Workout-Longevity/dp/0956310109*

Fig. 1.7 - (Page 14): Toothpick bridge images
*Image Credit: Author's Own*

Fig. 1.8 - (Page 15): Dr. Edward Dowdye picture
*Image Credit: SkyScholar YouTube - Can Stars BEND Light?*
*https://www.youtube.com/watch?v=B_ixkOI4k8c*

## Chapter 2

Fig. 2.10 (Page 43): Refractive index/ velocity of light shift diagram
*Image Credit: Author's Own*

Fig. 2.11 (Page 46): Emission from moving train diagram
*Image Credit: Author's Own*

Fig. 2.12 (Page 48): Relativity built upon haphazard foundation
*Image Credit: Author's Own*

Fig. 2.13 (Page 48): Analogy of building upon a foundation of inevitable failure
*Image Credit: Author's Own*

Fig. 2.14 (Page 49): Closed mind
*Image Credit: Author's Own*

Fig. 2.15 (Page 50): Demonstrating a wave-front with a stationary hose
*Image Credit: Author's Own*

Fig. 2.16 (Page 50): Demonstrating a wave-end with a stationary hose
*Image Credit: Author's Own*

Fig. 2.17 (Page 51): Demonstrating a wave-front with a moving hose
*Image Credit: Author's Own*

Fig. 2.18 (Page 51): Analogy of a moving laser pointer to Fig. 2.17
*Image Credit: Author's Own*

Fig. 2.19 (Page 52): Still images from my animation showing a stationary star sending out its first burst of light and the time it takes for that convex wave-front to reach Earth.
*Image Credit: Author's Own*

Fig. 2.20 (Page 52): Still images from my animation showing the stationary star's light ceasing and the time it takes for that concave wave-end to reach Earth.
*Image Credit: Author's Own*

Fig. 2.21 (Page 53): AI image of Albert Fly-nstein
*Image Credit: Author's Own*

Chapter 3

Fig. 3.11 (Page 83): Diagram I made of sine wave quadrants, properly
*Image Credit: Author's Own*

Fig. 3.12 (Page 89): Side-by-side comparison of the quantum Schrödinger equation vs modifications under Galilean variance/ Dr. Dowdye's Extinction Shift Principle (ESP)
*Image Credit: Author's Own*

Fig. 3.13 (Page 95): Cult of Einstein at Bohemian Grove. Notice Moloch in the background
*Image Credit: Author's Own*

Fig. 3.14 (Page 95): Einstein laying lifeless in bed wearing the Heaven's Gate cult outfit
*Image Credit: Author's Own*


## Chapter 4

Fig. 4.1 (Page 100): Screenshot from interactive page I made with control sliders to adjust the factors of the rotating mirror with laser experiment to show redshift solely as a difference of relative velocity and not dependent upon acceleration or direction.
*Image Credit: Author's Own*

Fig. 4.2 (Page 101): Chart of factors for the single rotating mirror/ laser redshift scenario.
*Image Credit: Author's Own*

Fig. 4.3 (Page 102): Chart of factor for the double rotating mirror/ laser redshift scenario
*Image Credit: Author's Own*

Fig. 4.4 (Page 105): Side-by-side comparison between the relativity/Lorentz invariance interpretation of the rotating mirror and laser experiment versus Galilean variance
*Image Credit: Author's Own*

Fig. 4.5 (Page 108): Diagram explaining the William de Sitter's double star misnomer
*Image Credit: Author's Own*

Fig. 4.6 (Page 109): Screenshot of a video I made showing an interactive page with control sliders demonstrating a change in the velocity of light between primary, secondary and tertiary sine waves. The sine waves are always in phase and automatically adjust.
*Image Credit: Author's Own*

Fig. 4.7 (Page 112): Diagram of Petr Beckman and Peter Mandics Lloyd Mirror experiment
*Image Credit: Author's Own*

Fig. 4.8 (Page 113): Beckmann & Mandic's interferometer assembly
*Image Credit: Petr Beckmann and Peter Mandics - Test of the Constancy of the Velocity of Electro- magnetic Radiation in High Vacuum - RADIO SCIENCE Journal of Research NBS/USNC-URSI Vol. 69D, No.4, April 1965*
*https://nvlpubs.nist.gov/nistpubs/jres/69D/jresv69Dn4p623_A1b.pdf*

Fig. 5.11 (Page 125): The same glass prism but pivoted an additional few unspecified degrees but viewed from a top-down angled perspective.
*Image Credit: Author's Own*

Fig. 5.12 (Page 125): The same glass prism pivoted an additional few unspecified degrees but from a top-down/arial view.
*Image Credit: Author's Own*

Fig. 5.13 (Page 128): Multiple images
(a) Ladybug with full spectrum of colors hitting the elytra but only red going into the eye. I modified the image, putting a red circle with slash through it to denote it is incorrect.
*Image Credit: https://betterthanpi.com/the-physics-of-light-colour/*
(b) Ladybug with full spectrum of colors hitting the elytra with all colors except red being re-emitting into the eye. I modified the image
*Image Credit: https://betterthanpi.com/the-physics-of-light-colour/*
(c) Negative space lettering.
*Image Credit: https://www.designspiration.com/save/2598016736833/*

Fig. 5.14 (Page 129): AI image of a morpho butterfly wing compared to a blue laser
*Image Credit: Author's Own*

Fig. 5.15 (Page 129): AI images of the microscopic scales and ridges of the morpho wings.
*Image Credit: Author's Own*

Fig. 5.16 (Page 130): Multiple images
(a) Green leaf with full spectrum of colors hitting the leaf but only green entering the eye.
*Image Credit: http://www.mstworkbooks.co.za/natural-sciences/gr8/images/gr8ec04-gd-0027.png*
(b) Same green leaf with full spectrum of colors hitting the leaf but with all colors except green being re-emitted into the eye.
*Image Credit: http://www.mstworkbooks.co.za/natural-sciences/gr8/images/gr8ec04-gd-0027.png*

Fig. 5.17 (Page 132): Johann Goethe and Isaac Newton parody of the "epic handshake" meme originating from the movie *Predator* with Arnold Schwarzenegger & Carl Weathers.
*Image Credit: Author's Own*

Fig. 5.18 (Page 133): Comparison of direct and indirect sunlight.
*Image Credit: Author's Own*

Fig. 5.19 (Page 134): The wheel of colors showing 180° orientation of the Newtonian primary colors to the Goethenian colors.
*Image Credit: WRMF.ca (Transferred) - Kevin McMillan*
*https://wrmf.ca/posts/the-hsl-color-model/*

Fig. 5.20 (Page 135): Einstein's shadow exposes him as a clown.
*Image Credit: Author's Own*

Fig. 5.21 (Page 135): I situated 3 lamps equidistant so the red, green and blue light bulbs shine onto a rotating model of a central planet. The planet is magnetically levitated.
*Image Credit: Author's Own*

Fig. 5.22 (Page 136): The results of switching on and off the different combinations of lights
*Image Credit: Author's Own*

Fig. 5.23 (Page 136): Comparison of all the colored lights off and all the colored lights on.
*Image Credit: Author's Own*

Fig. 5.24 (Page 147): How relativity suggests the sun SHOULD bend light. Never observed.
*Image Credit: Author's Own*

Fig. 5.25 (Page 148): Dr. Edward Dowdye's attempted submission to peer review and the anonymous referee rejecting the paper based upon a misconception of microwaves.
*Image Credit: Dr. Edward Dowdye*
*https://web.archive.org/web/20220122163149/http://www.extinctionshift.com/*

Fig. 5.26 (Page 150): Multiple images
Side-by-side comparisons between the light bending rule of GR versus actual observations as presented by Dr. Edward Dowdye on SkyScholar's YouTube channel.
*Image Credit: From video titled "Can Stars BEND LIGHT? General Relativity and Gravity with Dr. Edward Dowdye!" Screenshots taken from 1 min 10 seconds - 1 min 20 sec. And also 1:36 and 6:59.* https://www.youtube.com/watch?v=B_ixkOI4k8c

Fig. 5.27 (Page 151): Multiple images
(a) Straw in a glass of water demonstrating refracting only at the surface of the water.
*Image Credit: Photographer unknown*
(b) One of Eddington's May 29, 1919, solar eclipse photos.
*Image Credit: Photograph by Sir Arthur Eddington, taken during the solar eclipse on May 29, 1919. Presented in his 1920 paper announcing the successful test of general relativity. Source: Royal Society of London*

Fig. 5.28 (Page 153): How relativity suggests the sun SHOULD bend light. Never observed. I crossed out all the equations that do not apply, highlighting the equations only at the limb.
*Image Credit: Author's Own*

Fig. 5.29 (Page 154): Actual observations and data collected for the deflection of light only at the surface of the sun.
*Image Credit: Author's Own*

Fig. 5.30 (Page 155): How relativity says a black hole SHOULD bend light. Never observed.
*Image Credit: Author's Own*

Fig. 5.31 (Page 156): How relativity says a black hole SHOULD bend light. Never observed. I crossed out all the equations that do not apply, The only equations that apply when there is deflection are the equations at the surface of a body.
*Image Credit: Author's Own*

Fig. 5.32 (Page 157): Actual observations of Sagittarius A showing no deflection at all. Even at the alleged surface. This violates GR rule that SHOULD bend light.
*Image Credit: Author's Own*

Fig. 5.33 (Page 159): A slide from Dr. Robitaille's presentation called "The Sun on Trial".
*Image Credit: Screenshot from* https://www.youtube.com/watch?v=9TOKo7Ik9f8 ????

Fig. 5.34 (Page 161): Herschel image of Abell 2218
*Image Credit: Abell 2218 NASA. A. Fruchter & The ERO Team (STScI) • STScI-PRC00-08 (HST • WFPC2) Image captured in 1999*

Fig. 5.35 (Page 162): Multiple images
(a) NASA image of Abell 2218 (top)
*Image Credit: Abell 2218 NASA. A. Fruchter & The ERO Team (STScI) • STScI-PRC00-08 (HST • WFPC2) Image captured in 1999*
(b) Simulation of lensing of Abell 2218
*Image Credit: Bhuvnesh Jain, Uros Seljak, and Simon White ????*

Fig. 5.36 (Page 163): Abell 2218 by NASA
*Image Credit: Abell 2218 NASA. A. Fruchter & The ERO Team (STScI) • STScI-PRC00-08 (HST • WFPC2) Image captured in 1999*

Fig. 5.37 (Page 163): Abell 2218 Cluster by NASA
*Image Credit: Photograph by W.Couch & R.Ellis NASA uploaded on May 10th, 2013*

Fig. 5.38 (Page 163): I took the 2 images from Fig. 5.36 and 5.37 and overlapped them. I rotated Fig. 5.37 to show it's just the same image pawned off as a different one.
*Image Credit: Abell 2218 NASA. A. Fruchter & The ERO Team (STScI) • STScI-PRC00-08 (HST • WFPC2) Captured in 1999/Photographed by W.Couch & R.Ellis NASA uploaded May 10th 2013*

Fig. 5.39 (Page 165): I took the 2 images from Fig. 5336 and 5.37 and added them in an AI parody of the meme "They're the same picture" from the show The Office.
*Image Credit: Abell 2218 NASA. A. Fruchter & The ERO Team (STScI) • STScI-PRC00-08 (HST • WFPC2) Captured in 1999/Photograph by W.Couch/R.Ellis NASA uploaded May 10, 2013*

Fig. 5.40 (Page 165): The other half of the "They're the same picture" meme with an AI rendition of Jenna Fischer's character, Pam.
*Image Credit: Author's Own*

Fig. 5.41 (Page 167): Spiral galaxy pair NGC 3314 by NASA
*Image Credit: PRC00-14 Space Telescope Science Institute - NASA and the Hubble Heritage Team (STScI/AURA)*

Fig. 5.42 (Page 168): Multiple images
(a) G2 gas cloud projected computer simulation (top)
*Image Credit: ESO/S. Gillessen/MPE/Marc Schartmann/L. Calçada*
https://www.eso.org/public/videos/eso1332a/
https://buffalonews.com/opinion/snack-time-in-the-cosmos-astronomers-hope-to-view-giant-black-hole-consuming-gas-cloud/article_c1f6781b-da8a-5d5c-ad60-2630faa26105.html
(b) Comparison to the actual data collected by satellites of the G2 gas cloud (bottom)
*Image Credit: An image from W. M. Keck Observatory near infrared data shows that G2 survived its closest approach to the black hole in northern summer 2014. The green circle just to its right depicts the location of the invisible supermassive black hole.*
*Image via Keck Observatory*
*https://earthsky.org/space/how-g2-survived-the-black-hole-at-our-milky-ways-heart/*
*https://www.syfy.com/syfy-wire/bad-astronomy-dust-cloud-g2-is-actually-three-stars-with-clouds-around-them*

Fig. 5.43 (Page 169): Isaac Newton waking up in his coffin like the Undertaker from the WWE hearing a church bell, unbeknownst to Einstein.
*Image Credit: Author's Own*

Fig. 5.44 (Page 171): A man demonstrating the deflection of charged particle in a Crooke's tube by holding a magnet close to the tube. The green light shows the bent path traced.
*Image Credit: Mateus Silva Rego*
*https://en.wikipedia.org/wiki/Crookes_tube#/media/File:Ampola_de_Crookes.gif*

# Chapter 6

Fig. 6.1 (Page 174): Isaac Newton sits under a tree and gets hit in the head by an apple.
*Image Credit: Author's Own*

Fig. 6.2 (Page 176): Multiple images
A collection of 11 images depicting space-time curvature as per relativity.
(a) Interstellar black hole (top left)
*Image Credit: The supermassive black hole Gargantua from the 2014 sci-fi blockbuster "Interstellar." Paramount Pictures*
(b) Graphical depiction of 4D space-time bending around Earth. (top middle left)
*Image Credit: Image credit: Christopher Vitale of Networkologies and the Pratt Institute*
(c) Depiction of relativistic light bending of light near the Sun. (top middle right)
*Image Credit: Posted by AshrafSiddiqui on May 2007*
*https://cr4.globalspec.com/thread/112638/Gravitational-Lensing*

(d) Light bending by the Sun according to General Relativity
*Image Credit: Daniel Kiteck*
*https://www.pinterest.com/pin/general-relativity-predicts-the-gravitational-bending-of-light-by-massive-bodies--41447259040136307/*
(e) Depiction of a black hole computer generated.
*Image Credit: Artist unknown*
(f) Computer model of flat space-time then curving into a well around a black hole
*Image Credit: Artist unknown*
(g) Depiction of light bending near the Sun due to relativistic space-time curvature.
*Image Credit: https://respuestas.me/q/como-sabemos-que-la-curvatura-de-la-luz-alrededor-de-las-estrellas-se-debe-62368774548*
(h) Depiction of how light is supposed to bend and curve according to relativity.
*Image Credit: NASA, ESA, L. Calcada*
(i) Another image of light traveling straight but bending in the presence of Earth's gravity.
*Image Credit: Artist unknown*
(j) Depiction of flat space-time bending in the presence of mass.
*Image Credit: NASA, ESA, L. Calcada*
*https://www.esa.int/ESA_Multimedia/Images/2015/09/Spacetime_curvature*
(k) Computer image of
*Image Credit: "Public Domain"*
*https://www.businessinsider.com/gravitational-wave-ligo-detection-importance-2016-1*

Fig. 6.3 (Page 176): Einstein's psychiatrist asking him if space-time curvature is in the room
*Image Credit: Author's Own*

Fig. 6.4 (Page 179): Multiple images from my slide presentation on Galilean variance.
(a) John Stuart Reid's model of sound in 3 dimensions. (top left)
*Image Credit: John Stuart Reid*
(b) Spherical wavefronts - physics video YouTube (top right)
*Image Credit: CBSE Class 12 Physics, Wave Optics – 1, Huygens Principle - Uploaded by YouTube user BestForNeet*
*https://www.youtube.com/watch?v=gliLaaeZHwg*
(c) Dr. Edward Dowdye's Extinction Shift examples/visuals from his site. (middle left)
*Image Credit:*
https://web.archive.org/web/20160331150730/http://www.extinctionshift.com/details06.htm
(d) Dr. Edward Dowdye's Extinction Shift examples/visuals from his site. (middle right)
*Image Credit:*
*https://web.archive.org/web/20160331145744/http://www.extinctionshift.com/details05.htm*
(e) Spherical wavefronts - physics video YouTube (bottom left)
*Image Credit: CBSE Class 12 Physics, Wave Optics – 1, Huygens Principle - Uploaded by YouTube user BestForNeet*
*https://www.youtube.com/watch?v=gliLaaeZHwg*
(f) Dr. Edward Dowdye's ExtinctionShift.com site (bottom middle)
*Image Credit:*
*https://web.archive.org/web/20160331145545/http://www.extinctionshift.com/details03.htm*

(g) Computer animated depiction of a constant emission of spherical wavefronts from a point source. (bottom right-top)
*Image Credit: CBSE Class 12 Physics, Wave Optics – 1, Huygens Principle - Uploaded by YouTube user BestForNeet*
*https://www.youtube.com/watch?v=gliLaaeZHwg*
(h) A different Computer animated depiction of a constant emission of spherical wavefronts from a point source. (bottom right-bottom)
*Image Credit: CBSE Class 12 Physics, Wave Optics – 1, Huygens Principle - Uploaded by YouTube user BestForNeet*
*https://www.youtube.com/watch?v=gliLaaeZHwg*

Fig. 6.5 (Page 181): Emission and re-emission of gravity diagram
*Image Credit: Author's Own*

Fig. 6.6 (Page 184): Side-by-side diagram gravity equations between Lorentz invariance (relativity) versus Galilean variance.
*Image Credit: Author's Own*

Fig. 6.7 (Page 186): Atlas spinning the Earth on his finger with ease rather than struggling to lift the weight on his back.
*Image Credit: Author's Own*

# Glossary of Mathematical Symbols

$\approx$   (about equal to)

$\neq$   (not equal to)

$\pm$   (plus or minus)

$^2$   (squared)

$^\circ$   (degrees)

$\infty$   (infinity)

$\propto$   (Proportional to)

$\sqrt{}$   (square root)

$\rightarrow$   (function mapping and vectors) Used to show something changing to something else

$\int$   (Integral symbol) Used in calculus for finding areas, volumes and quantities

$\Sigma$   (sigma Uppercase) Used for Sums

$\vec{E}$   (electric field vector) Used to describe the electric field for magnitude and direction

$\in$   (element of/ belongs to)

$\epsilon$   (Technical epsilon for "per"/ signifies a small positive quantity, especially in limits

$\partial$   (Partial derivative symbol) Used to keep one variable constant in relation to others

$\nabla$   (Nabla or Del operator) Used as a gradient of divergence in vector calculus

$\Delta$   (delta uppercase) Rate of change

$\alpha$   (alpha lowercase) Used for alpha particles, angles and coefficients

$\beta$   (beta lowercase) Used for beta decay and coefficients

$\gamma$   (gamma lowercase) Used for Lorentz factor and gamma rays

$\delta$   (delta lowercase) A small quantity of change

$\varepsilon$   (epsilon lowercase) Used to describe quantity in limits

$\zeta$   (zeta lowercase) Used for dampening, prime number distribution, angles fluid dynamics

$\theta$   (theta lowercase) Used to describe angles

$\kappa$   (kappa lowercase) Used for the spring constant or dielectric constant

$\lambda$   (lambda lowercase) Used for wavelength

$\mu$   (mu lowercase) Used for micro and coefficient for friction

$\nu$   (nu lowercase) Used for frequency in wave mechanics

$\xi$   (xi lowercase) Used for non-linear dynamics

$o$   (omicron lowercase) Used for infinitesimally small quantities

$\pi$   (pi lowercase) Used for ratio of circumference of circle to its diameter

$\rho$   (Rho lowercase) Used for Density

σ  (Sigma lowercase)  Used for standard deviation

τ  (Tau lowercase) Used for time constants in decay, coefficients

Υ  (Upsilon uppercase) Used for mesons and particle physics

φ  (phi lower case)  Used for phase shift, golden ratio

χ  (chi lowercase) Used for distribution and electromagnetism

ψ  (psi lowercase) Used for wavefunctions in quantum mechanics

ω  (omega lowercase)  Used for angular frequency

Γ  (Gamma uppercase)  Used for gamma function

Θ  (Theta uppercase) Used for temperature in thermodynamics and statistics

Λ  (Lambda uppercase) Used for linear algebra and wavelengths

Φ  (Phi uppercase) Used for magnetic flux, electromagnetism, golden ratio

Ψ  (Psi uppercase) Used for wavefunctions in quantum mechanics

Ω  (Omega uppercase) Used for ohms

℧  (inverse ohm) Used for electrical conductance

Å  (Angstrom)  Used as a unit of measurement 0.1 nanometers

$\hbar$  (H-bar)  Used as reduced Planck constant in quantum mechanics

$m$  Used for mass in general

$m_{eff}$  Used for Effective mass (the Galilean variant equivalent to relativistic mass)

$A$  Used for Amplitude

$f$  Used for frequency

$t$  Used for time

$r$  Used for radius of a circle or sphere/ geometry

$c$  Used for the velocity of light constant

$E$  Used for Energy

$M$  Used for mass as a whole system

$R$  Used for radius specific to curvature like a celestial body or multiple radii

$G$  Used for gravitational constant

$g$  Used for the acceleration due to gravity

$'$  (Apostrophe/ Prime) Used to denote a modification or different understanding to an equation or used to denote a different frame of reference, or derivative of a function

$Sin$  Used for trigonometric functions

$Cos$  Used for trigonometric functions

$n$  Used for refractive index

h  Used for Planck's constant

$V$  Used for volume

$k$  Used for a constant or coefficient

J   Used for Joules

X   Used as a multiplication symbol

**B**   Used for magnetic field in Maxwell/Heaviside/ Ampére's equations

**F**   Used for Force

$q$   Used for electric charge

**V**   Used for velocity vector of the charge

$[x,p]$   Used for position operator, and momentum operator

$i$   Used for the square root of -1

$H$   Used for Hamiltonian, classical particle physics and classical wave mechanics

$v_{eff}$   Used for Effective velocity. A Galilean variant equivalent to relativistic velocity

$u$   Used to signify uniformity

$Z$   Used for Zahlen/ numbers meaning all set of integers

$S$   Used to represent the "source" of an emission or source of light/ origin of a primary

x, y, z  Used for the three orthogonal axis in 3D Euclidean space

$e$   Used for Euler's number

$\therefore$   Used to represent "therefore"

app/ rec  Used to represent "approaching" and "receding" away

$I$   Used for Intensity

$T$   Used for Temperature

$M_c$ : Used for critical mass

$f(p, P)$ : The function relating to the density and pressure and critical mass.

$\Gamma_{abs}$ : The rate of absorption of gravitons by fermions making up surrounding matter.

$\sigma$ : Describes the likelihood of fermions to absorb gravitons.

$n_f$: The density/concentration and number of fermions available to absorb gravitons.

$\Phi_{primary}$: The emissions of primary gravitons at the boundary of a coherent state.

$\eta$ : The efficiency of re-emission of gravitons describing fractions of graviton secondaries.

$\psi'_{primary}(\mathbf{r}, t)$: Modified wave function of primary gravitons over space & time. (ESP)

$\psi'_{secondary}(\mathbf{r}, t)$: Modified wave function - secondary gravitons over space & time (ESP)

$\sum_{\cdot n} a_n$: The sum of all terms representing the sum of influences from graviton emissions.

$K$: Proportionality constant for the relationship of density and pressure.

$P_c$: The critical pressure threshold required for a phase transition.

$p_c$: The critical density threshold at which the phase transition occurs.

$Q$ is the energy in the form of heat measured in joules.

# References, Publications & Citations

Newton, Isaac. *Philosophiæ Naturalis Principia Mathematica*. London: Joseph Streater, 1687.

Edward H. Dowdye, Jr. *Discourses & Mathematical Illustrations pertaining to the Extinction Shift Principle*, 2012, ISBN 0-936-4471-5-7
https://web.archive.org/web/20220122163149/http://www.extinctionshift.com/

Beckmann, Petr. *Einstein Plus Two*. 1987. ISBN 0-911762-39-6.

Hatch, Ron. *Escape from Einstein*. 1992. ISBN 0-9632113-0-7.
Talbert, Philip H. *The Half-Life of a Nuclear Battery*. 2008. ISBN 978-0-615-23375-8.

Michalets, David. *Cosmology Transition*. 2008.

Michelson, Albert A. *"Effect of Reflection from a Moving Mirror on the Velocity of Light."* *Astrophysical Journal*, vol. 38, no. 5, 1913, pp. 421-426.

Beckmann, Petr and Mandics, Peter, Test of the Constancy of the Velocity of Electromagnetic Radiation in High Vacuum, Radio Sci. J. Res. NBS/USNC.URSI 69D, No. 4, 623 - 628 (1965)
Babcock, G.C., and Bergman, T.G., Determination of the Constancy of the Speed of Light, J. Opt. Soc. Am. 54 - 147 - 151 (1964)

Koechner, Walter. *Optical Sciences Solid State Engineering*. 4th ed. ISBN 3-540-60237-2

Michelson, A.A., Effect of Reflection from a Moving Mirror on the Velocity of Light, Astrophysics. J., vol. 37, 190 - 193 (1913)

Michelson, Albert A., and Edward W. Morley. "On the Relative Motion of the Earth and the Luminiferous Ether." *American Journal of Science*, vol. 34, no. 203, 1887, pp. 333-345.

Unzicker, Alexander. *This Higgs Fake*. 2013. ISBN 1492176249
Searl, John. *Law of Squares Book Series, Books 1, 1A, 1B, 1C, Book 4*, 1960s

Dowdye, E.H., Jr. "The Solar Plasma Limb Found to Deflect Microwaves from Extra Galactic Radio Sources at Lowest Impact Parameter Corresponding to Solar Plasma Limb." 2018. Web. https://web.archive.org/web/20220124082433/http://www.extinctionshift.com/PaperSubmittedToAstronomischeNachrichten.pdf

Dowdye, E.H., Jr. "Extinction Shift Principle: A Pure Classical Alternative to General and Special Relativity." *Physics Essays*, vol. 20, 56, 2007, pp. 1-11. DOI: 10.4006/1.3073809. http://quantumrealism.net/wp-content/uploads/2019/03/1c1q1qcoh_33307.pdf

Li, Aigen. 2009, "Optical Properties of Dust", Springer, Chapter 6, pp. 167-188

Li, Aigen, Drain, B.T., 2002, "Are Silicon Nanoparticles an Interstellar Dust Component?", The Astrophysical Journal, 564, 803

Draine, B.T., 2003, "Scattering by Interstellar Dust Grains. I. Optical and Ultraviolet", The Astrophysical Journal, 598: 1017-1025

Kozasa, T., Blum, J., Okamoto, H., Mukai, T., 1993, "Optical properties of dust Aggregates: II. Angular dependence of scattered light", Astronomy and Astrophysics, 276, 278-288

Dowdye, E.H., Jr. "Time Resolved Images from the Center of the Galaxy Appear to Counter General Relativity." *Astronomische Nachrichten*, vol. 328, no. 2, Feb. 2007, pp. 186-191. Web. http://www3.interscience.wiley.com/search/allsearch.

Dowdye, Edward H., Jr. "The Photon and Its Measurability." SPIE, 2005.

Dowdye, E.H., Jr. "Extinction Shift Principle: A Pure Classical Alternative to General and Special Relativity." *Physics Essays*, vol. 20, no. 56, 2007, pp. 1-11. DOI: 10.4006/1.3073809.

Lebach, D. E. et al., "Measurement of the Solar Gravitational Deflection of Radio Waves Using Very-Long-Baseline Interferometry", Phys. Rev. Lett, 75 (1995), pp. 1439-1442

"Are the Conventional Concepts of Gravitational Lensing Adhering to the Observational Evidence and Mathematical Physics Fundamentals?" *Infinite Energy*, vol. 15, no. 88, 2009, http://www.infinite-energy.com/iemagazine/issue88/index.html.

Counselman, C.C., et al. "Solar Gravitational Deflection of Radio Waves Measured by Very-Long-Baseline Interferometry." *Physical Review Letters*, vol. 33, 1974, pp. 1621-1623.

Fomalont, E.B., et al. "Measurements of the Solar Gravitational Deflection of Radio Waves in Agreement with General Relativity." *Physical Review Letters*, vol. 36, 1976, pp. 1475-1478.

Carter, James. "The True Direction of Gravitational Force." *Proceedings of the Natural Philosophy Alliance*, 18th Conference of the NPA, 6-9 July 2011, University of Maryland, College Park, USA, vol. 8, pp. 107-109.

Dowdye, E.H., Jr. "Extinction Shift Principle: A Pure Classical Alternative to General and Special Relativity." *Physics Essays*, vol. 20, no. 56, 2007, pp. 1-11. DOI: 10.4006/1.3073809.

Neil Ashby, University of Colorado
https://web.archive.org/web/20010815033201/http://vishnu.nirvana.phys.psu.edu/mog/mog9/node9.html

Dana, Peter H. "Global Positioning System (GPS) Time Dissemination for Real-Time Applications." *Department of Geography, University of Texas at Austin*, 2005

Simanek, Donald E. "Tidal Misconceptions." *Lock Haven University*, 2017. Web. https://web.archive.org/web/20170317074855/www.lhup.edu/~dsimanek/scenario/tides.htm

Ghanzanshahi, Shahin, and Edward H. Dowdye, Jr. "Shapiro Delay: A Frequency Dependent Transit Time Effect." *Proceedings of SPIE*, vol. 8121, 2011. California State University, Fullerton, and Pure Classical Physics Research

Epstein, Joseph. "The Shapiro Experiment." Web. https://www.relativity.li/uploads/pdf/English/Epstein_en.pdf

Chandler, J. F., et al. "Solar-System Dynamics and Tests of General Relativity with Planetary Laser Ranging." *NASA*, 2000

Muhleman, Duane O., and John D. Anderson. "Solar Wind Electron Densities from Viking Dual-Frequency Radio Measurements." *The Astrophysical Journal*, vol. 247, 1981, pp. 1093-1101

Sears, Zemansky, Young and Freedman, University Physics, 10th Ed., Addison-Wesley, (2000) Change in Velocity of Light emitted by a Moving Source - J. James R. Sternberg

Babcock, G. C., and T. G. Bergman. "Determination of the Constancy of the Speed of Light" *Journal of the Optical Society of America* 1964

Anderson, John D., Paquale B. Esposito, Warren Martin, and Catherine L. Thornton. "Experiment Test of General Relativity Using Time-Delay Data from Mariner 6 and Mariner 7." *The Astrophysical Journal*, vol. 200, 1975, pp. 221-233

Muhleman, Duane O., Paquale B. Esposito, and John D. Anderson. "The Electron Density Profile of the Outer Corona and the Interplanetary Medium from Mariner-6 and Mariner-7 Time-Delay Measurements." *The Astrophysical Journal*, vol. 211, 1977, pp. 943-957

Born, Max, and Emil Wolf. *Principles of Optics*. Pergamon Press, 1975, pp. 71, 100-104

Jackson, John David. *Classical Electrodynamics*. John Wiley & Sons, Inc., 1975, pp. 512-515

Beckmann, Petr & Mandics, Peter, Experiment on the Constancy of the Velocity of Electromagnetic Radiation, Radio Sci. J Res NBS/USNC/URSI 68D, No 12, 1265-1268 (1964)

Gerber, Paul. "Die Fortpflanzungsgeschwindigkeit der Gravitation." *Malm. u. Phys.*, vol. 43, 1898, pp. 93-104. Reprint in *Annalen der Physik*, vol. 52, 1917, pp. 415-444

Einstein, Albert. "The Foundation of the General Theory of Relativity." Annalen der Physik, vol. 49, 1916, pp. 769-822

B. H. D. Johnson. "The Historical Development of the Einstein Field Equations." General Relativity and Gravitation, vol. 29, 1997, pp. 611-617

Rotz, Fred B, New Test of the Velocity of Light, Physics Letters 7 No 4 252-254 (1963)

James, J F, & Sternberg, R.S., Change in Velocity of Light Emitted by a Moving Source, Nature 197, 1192 (1963)

Dowdye, E.H., Extinction Shift; an Alternative to the Doppler Shift Theory, (Copyrighted work published) (1983)

Dowdye, E.H., Extinction Shift Principle Illustrated; Some Classical Alternatives Equivalent to Special and General Relativistic Principles, (Copyrighted work published) (1991)

Hulse, R.A, and Taylor, J.H., Science250, 770 (1990); Hulse, R.A. and Taylor, J.H., Astrophys. J. 195, L51 (1975)

Redondi, Pietro, Galileo: Heretic (Galileo Eretico), Princeton U. Press, Princeton, NJ (1987)

G Arfken, Hans Weber, Mathematical Methods for Physicist. Academic Press p 77-79 (1995)

R. Serway, J. Faugh, College Physics, Philadelphia - New York, Saunders College Publishing, 5, 157 - 166 (1999)
John David Jackson, Classical Electrodynamics, 3rd. ed., John Wiley & Sons, Inc., pp. 27-29 (1998) Resnick, Walker, Fundamentals of Physics, 5E, Extended, Wiley (1997)

Lorentz, H.A. "Electromagnetic Phenomena in a System Moving with Any Velocity Less Than That of Light." *Proceedings of the Academy of Sciences Amsterdam*, vol. 6, 1904. Reprinted in *The Principle of Relativity*, Dover Publications, New York, 1952. See also *The Theory of Electrons*, 2nd ed., reprinted by Dover Publications, New York, 1952

Herouni, P. "Measured Parameters of Large Antenna of ROT-54/2.6 Tell About Absence of Big Bang." *Journal of Astrophysics: Reports*, vol. 107, no. 1, 2007, pp. 73-78. National Academy of Sciences of Armenia. http://rnas.asj-oa.am/2542/1/73.pdf

Herouni, P. "About Self Noises of Radio-Optical Telescope ROT-54/2.6 Antenna." *Journal of Applied Electromagnetism*, vol. 2, no. 1, 1999, pp. 51-57. Athens. http://jae.ece.ntua.gr/archive/1999/vol2no2_June1999.zip

16 Peer Reviewed Papers by Dr. Dowdye. Hosted by NASA and Harvard. *ADS* https://ui.adsabs.harvard.edu/search/q=dowdye&sort=date%20desc%2C%20bibcode%20desc&p_=0

Dowdye, Edward. "Double Slit Re-Emission Explanation." *ExtinctionShift.com* Web. https://web.archive.org/web/20210906211144/http://www.extinctionshift.com/DOUBLE_SLIT_Topic/DOUBLE_SLIT_topic.htm

Robitaille P.-M. and Crothers S.J. "The Theory of Heat Radiation" Revisited: A Commentary on the Validity of Kirchhoff's Law of Thermal Emission and Max Planck's Claim of Universality. Prog. Phys., 2015, v. 11, no. 2, 120–132

Robitaille, P.M. "WMAP: A Radiological Analysis." *Prog. in Physics*, vol 3, no. 1, 2007, p. 3-18

Robitaille, P.M. "Water, Hydrogen Bonding and the Microwave Background." *Progress in Physics*, vol. 5, no. 2, 2009, pp. L5-L8

Robitaille, P.M. "COBE: A Radiological Analysis." *Prog in Physics*, vol 5, no 4, 2009, p 17-42

Robitaille, P.M. "The Planck Satellite LFI and the Microwave Background: Importance of the 4 K Reference Targets." *Progress in Physics*, vol. 6, no. 3, 2010, pp. 11-1

Robitaille, P.M. "Kirchhoff's Law of Thermal Emission: 150 Years." *Progress in Physics*, vol. 4, 2009, pp. 3-13

Robitaille, P.M. "Blackbody Radiation and the Carbon Particle." *Progress in Physics*, vol. 3, 2008, pp. 36-55

Robitaille, P.M. "The Theory of Heat Radiation Revisited: A Commentary on the Validity of Kirchhoff's Law of Thermal Emission and Max Planck's Claim of Universality." *Progress in Physics*, vol. 12, no. 3, 2016, pp. 184-203

Robitaille, P.M. "Stellar Opacity: The Achilles' Heel of the Gaseous Sun." *Progress in Physics*, vol. 3, 2011, pp. 93-99

Robitaille, P.M. "Forty Lines of Evidence for Condensed Matter - The Sun on Trial: Liquid Metallic Hydrogen as a Solar Building Block." *Progress in Physics*, 2013, pp. 90-142

Arp, Halton. *Seeing Red: Redshifts, Cosmology, and Academic Science*. 1998

Steinmetz, Charles Proteus. *Theory and Calculation of Alternating Current Phenomena*. W.J. Johnston Company, 1907

Steinmetz, Charles Proteus. *General Lectures on Electrical Engineering*. 2nd ed., compiled and edited by Joseph Le Roy Hayden, McGraw-Hill Book Company, 1911

Steinmetz, Charles Proteus. *Elementary Lectures on Electric Discharges, Waves and Impulses, and Other Transients*. McGraw-Hill Book Company, 1911

Steinmetz, Charles Proteus. *Radiation, Light and Illumination: A Series of Engineering Lectures Delivered at Union College*. Compiled and edited by Joseph Le Roy Hayden, 3rd ed., McGraw-Hill Book Company, 1918

Steinmetz, Charles Proteus. *Theoretical Elements of Electrical Engineering*. 3rd ed., McGraw-Hill Book Company, 1918

Steinmetz, Charles Proteus. *Engineering Mathematics: A Series of Lectures Delivered at Union College*. McGraw-Hill Book Company, 1911

Russell, Walter Bowman. *A New Concept of the Universe*. 1948

Russell, Walter Bowman. *The Russell Genero-Radiative Concept of Cyclical Theory of Continuous Motion*. 1954

Leedskalnin, Ed. *Magnetic Current*. 1945

Davis, Albert Roy/ Walter C. Rawls, Jr *The Magnetic Blueprint of Life*. Exposition Press, 1986

Davis, Albert Roy/ Walter C. Rawls, Jr. *Magnetism and Its Effects on the Living System*, 1976

Larson, Dewey B. *The Case Against the Nuclear Atom*. Reciprocal System Publishing, 1992

Pond, Dale, John Keely, Nikola Tesla, Edgar Cayce, *Universal Laws Never Before Revealed: Keely's Secrets: Understanding and Using the Science of Sympathetic Vibration*, 1999

Wright, Walter C. *Gravity Is a Push*. Carlton Press, 1975

Jenny, Hans. *Cymatics*. 1st & 2nd ed., 1967

Meyl, Konstantin. *Scalar Waves*. 2013

Michelson, A.A. "Relative Motion of Earth and Ether." *Philosophical Magazine*, vol. 8, pp. 716-719, 1904

Michelson, A.A. "Effect of Reflection from a Moving Mirror on the Velocity of Light." *Astrophysical Journal*, vol. 37, pp. 190-193, 1913

Michelson, A.A., and H.G. Gale. "The Effect of the Earth's Rotation on the Velocity of Light." *Astrophysical Journal*, vol. 61, pp. 137-145, 1925

Lerner, Eric. *The Big Bang Never Happened*. Vintage Books, 1992

Christopher Jon Bjerknes. *The Incorrigible Plagiarist: Albert Einstein* (2009)

Maurice B. Cooke. *Einstein Doesn't Work Here Anymore* (2006)

Christopher Jon Bjerknes. *Anticipations of Einstein in the General Theory of Relativity* (2005)

H.E. Retic. *The Einstein Hoax* (1998)

Page Truitt. *The Great Einstein Relativity Hoax: And Other Science Questions* (2010)

Roger Schlafly. *How Einstein Ruined Physics* (2007)

Milan P. Pavlovic. *Einstein's Theory of Relativity: Scientific Theory or Illusion* (2002)

Sigma, Rho. *Ether-Technology: A Rational Approach to Gravity Control* 1977

Dodson, J., and Poston, E. Tensor Geometry, 1990

McMahon, D. Relativity Demystified, 2006

Hawking, S., and Ellis, G.F.R. The Large Scale Structure of Space Time, 1973

Mould, R. Basic Relativity, 2000

Schutz, B. A Short Course in General Relativity, 1985

Dirac, P.A.M. General Relativity, 1975

Misner, C.W., Thorne, K.S., and Wheeler, J.A. Gravitation, 1973

d'Inverno, R. Introducing Einstein's Relativity, 1992

Weinberg, S. Gravitation and Cosmology, 1972

Wald, R.M. General Relativity, 1984

Hawking, S., and Penrose, R. The Singularity Theorems of Hawking and Penrose, 1996

Ludvicen, A. General relativity - A Geometric Approach, 2005

Hawking, S. A Brief History of Time, 1988

# Index

www.ingramcontent.com/pod-product-compliance
Lightning Source LLC
Chambersburg PA
CBHW081810200326

41597CB00023B/4205